Information Technology and Law Series

Volume 23

For further volumes:
http://www.springer.com/series/8857

Mark N. Gasson · Eleni Kosta
Diana M. Bowman
Editors

Human ICT Implants: Technical, Legal and Ethical Considerations

T·M·C·ASSER PRESS   Springer

Editors

Mark N. Gasson
School of Systems Engineering
University of Reading
Reading
Berkshire RG6 6AY
UK

Eleni Kosta
Interdisciplinary Centre for Law and
   ICT (ICRI)
KU Leuven
Sint Michielsstraat 6
3000 Leuven
Belgium

Diana M. Bowman
Risk Science Centre and the Department
   of Health Management and Policy
University of Michigan
Washington Heights 1420
Ann Arbor
MI 48109-2029
USA

ISSN 1570-2782
ISBN 978-90-6704-869-9   ISBN 978-90-6704-870-5   (eBook)
DOI 10.1007/978-90-6704-870-5

Library of Congress Control Number: 2012937224

© T.M.C. ASSER PRESS, The Hague, The Netherlands, and the author(s) 2012

Published by T.M.C. ASSER PRESS, The Hague, The Netherlands www.asserpress.nl
Produced and distributed for T.M.C. ASSER PRESS by Springer-Verlag Berlin Heidelberg

No part of this work may be reproduced, stored in a retrieval system, or transmitted in any form or by any means, electronic, mechanical, photocopying, microfilming, recording or otherwise, without written permission from the Publisher, with the exception of any material supplied specifically for the purpose of being entered and executed on a computer system, for exclusive use by the purchaser of the work. The use of general descriptive names, registered names, trademarks, etc. in this publication does not imply, even in the absence of a specific statement, that such names are exempt from the relevant protective laws and regulations and therefore free for general use.

Printed on acid-free paper

Springer is part of Springer Science+Business Media (www.springer.com)

Series Information

The *Information Technology & Law Series* was an initiative of IT*e*R, the National
programme for Information Technology and Law, which is a research programme
set up by the Dutch government and the Netherlands Organisation for Scientific
Research (NWO) in The Hague. Since 1995 IT*e*R has published all of its research
results in its own book series. In 2002 ITeR launched the present internationally
orientated and English language *Information Technology & Law Series*. This series
deals with the implications of information technology for legal systems and
institutions. It is not restricted to publishing ITeR's research results. Hence,
authors are invited and encouraged to submit their manuscripts for inclusion.
Manuscripts and related correspondence can be sent to the Series' Editorial Office,
which will also gladly provide more information concerning editorial standards
and procedures.

Editorial Office
eLaw&Leiden, Centre for Law in the Information Society
Leiden University
P.O. Box 9520
2300 RA Leiden
The Netherlands
Tel.: +31-71-527-7846
e-mail: ital@law.leidenuniv.nl

A.H.J. Schmidt, *Editor-in-Chief*
eLaw@Leiden, Centre for Law in the Information Society, Leiden University,
The Netherlands

Chr.A. Alberdingk Thijm, *Editor*
SOLV Advocaten, Amsterdam, The Netherlands

F.A.M. van der Klaauw-Koops, *Editor*
eLaw@Leiden, Centre for Law in the Information Society, Leiden University,
The Netherlands

Ph.E. van Tongeren, *Publishing Editor*
T.M.C. ASSER PRESS, The Hague, The Netherlands

Foreword

In March 2005, the European Group on Ethics in Science and New Technologies (EGE) to the European Commission, of which I was a member, issued an Opinion on "Ethical aspects of ICT implants in the human body" (EGE 2005). Since then many new, and in many cases exciting, developments in digital technology have taken place. These include, for example, affective computing, ambient intelligence, bioelectronics, cloud computing, neuro-electronics, future internet, artificial intelligence, human-machine symbiosis, quantum computing, robotics, and augmented reality. These advancements have given rise to further ethical and legal debates, especially in Europe (ETICA 2011). Concerns about human dignity, freedom, privacy, and data protection, freedom of research, identity and personality, autonomy and informed consent are no less important than questions of justice (fairness, equality, and solidarity) particularly regarding consumer protection, improvement and protection of health, as well as the application of the principles of proportionality, transparency, and precaution. In their "Ethical Evaluation" (Nagenborg and Capurro 2011, 4.8.2.1) remarked that,

> [I]n contrast to the possible use of ICT implants in AmI [Ambient Intelligence] [...] or in bio-electronics, neuro-electronic applications might raise greater ethical concerns, since neuro-electronics aims to provide a direct link between computer technology and the human brain as well as the nervous system in general. Especially the use of pre-conscious brain information-processing has to be seen as being in conflict with the protection of human dignity as highlighted in the "Ethical Analysis". It has to be recognized that (medical applications of) neuro-electronics might also contribute to the welfare of human life and therefore foster human dignity.

This ethical ambivalence should not be understood as though technology were *per se* neutral and all we needed was to distinguish between so-called good or bad uses. From a more basic perspective, it can be stated that no technology is neutral insofar as it *changes* the horizon of options we have had so far. In this sense, thinking about technology means thinking about such changes in ourselves and of our selves. This is highly relevant when dealing with ICT implants in the human body. There is an underlying Cartesian dualism in this reasoning insofar as changes in the human body cannot be separated from the question about *who* we are, which is the ethical perspective. From a (medical-) technical perspective, the human body is an information-processing device similar to other living systems that can be 'repaired' or even 'enhanced'. But, in fact, *my* body concerns primarily *who* I am and not only *what* I am. Who am I? I am not an isolated subject, but a self in interrelations with other selves: Family, friends, colleagues, etc. with whom I share a world. Who I am is a question of the social "interplay" with others in the world (Eldred 2008).

Today, this interplay is cast and decisively shaped by digital technology. Until recently, it has been argued, we were "bodies in technology" (Ihde 2002); now "technics" is becoming "embodied" (Ihde 2010). Both issues concern ourselves— and not just our bodies—in the world. From a whoness-related ethical perspective, ICT implants are not 'just' in *the* human body but in *my* human body. My body cannot be isolated from my self or from my being-with-others in the world. Once we are aware of this existential foundation of technology in general and of ICT implants in particular, we are able to ask ethical questions as pertaining not to an anonymous human body but to a singular and irreducible 'who' facing her past, present and future life, taking her risks and responsibilities toward herself no less than toward others in a shared world. Some of these questions were raised by the EGE in 2005 and they are today, I think, as relevant as they were a few years ago:

> How far can such implants be a threat to human autonomy, particularly when they are implanted in our brains? How far can such implants have irreversible impacts on the human body and/or on the human psyche and how can reversibility be preserved? [...] How far can ICT implants become a threat to privacy? [...] What lies behind the idea of an 'enhanced' human being? [...] How do we relate to persons with ICT implants that are connected online? How far should ICT implants remain invisible to an external observer? [...] How far can they be used in order to track human beings and in which cases should this be legally allowed? (EGE 2005, p. 24–25)

The EGE underscored that non-medical surveillance applications of human ICT implants "may only be permitted if the legislator considers that there is an urgent and justified necessity in a democratic society (Article 8 of the Human Rights Convention) and there are no less invasive methods." (EGE 2005, p. 34) I think that, due to new technological developments, this recommendation is today even more crucial than it was in 2005.

The present volume offers up-to-date, multifaceted contributions on the technical, legal, and ethical debate around human ICT implants by well-known academic experts. This debate should become not only a matter of professional thinking of ethicists, computer scientists, lawyers, the IT industry and policy-makers but, above all, of personal thinking and decision-making in everyday life.

Professor Rafael Capurro
International Center for Information Ethics (ICIE)
Distinguished Researcher in Information Ethics, School of
Information Studies, University of Wisconsin-Milwaukee,
Milwaukee, USA

References

Capurro R (2011) Toward a Comparative Theory of Agents. In AI & Society.http://www. springerlink.com/content/10494r4821528k87/

EGE (2005) Ethical aspects of ICT implants in the human body. Opinion No 20, (Rapporteurs: Stefano Rodotà and Rafael Capurro) http://ec.europa.eu/bepa/european-group-ethics/ publications/opinions/index_en.htm

Eldred M (2008) Social Ontology. Recasting Political Philosophy Through a Phenomenology of Whoness. Frankfurt am Main. Ontos Verlag. http://www.arte-fact.org/sclontlg.html

ETHICBOTS (2008) Emerging Technoethics of Human Interaction with Communication, Bionics, and Robotic Systems (European Union FP6 Project). http://ethicbots.na.infn.it/index. php

ETICA (2011) Ethical Issues of Emerging ICT Applications. (European Union FP7 Project) http://ethics.ccsr.cse.dmu.ac.uk/etica

Ihde D (2002) Bodies in Technology. University of Minnesota Press, Minneapolis

Ihde D (2010) Embodied Technics. Automatic Press, New York

Nagenborg M and Capurro R (2011) Deliverable 3.2.2 Ethical Evaluation. In: ETICA http://ethics.ccsr.cse.dmu.ac.uk/etica

Preface

This book would be but an idea without the generosity, commitment, and unwavering support of each of the contributing authors. As leaders in their respective fields, it is their insights, experience, and expertise that make this volume timely, thought provoking, and innovative. As the Editors of this book, we would like to express our appreciation to each of the contributors. The initial idea for the creation of this book was born from a report produced in the frame of the FP5 Network of Excellence 'FIDIS' (Future of Identity in the Information Society), and therefore, we would like to thank the partners of the FIDIS NoE. Special thanks go to Dr. Carmela Troncoso, Mr. Hans Hedbom, and Dr. Vassiliki Andronikou.

The Editors would also like to thank their respective universities, without whose support projects such as this one could not be undertaken, as well as their professional colleagues and their families. Each of the Editors brought their own skills to this book but both Eleni Kosta and Mark Gasson would particularly like to thank Diana Bowman for her enthusiasm in turning this project into reality. Finally, the series Editors, especially Franke van der Klaauw, and the staff at T.M.C. Asser Press deserve sincere thanks for their professionalism and their unstinting support.

February 2012

Dr. Mark Gasson
Dr. Eleni Kosta
Dr. Diana Bowman

Contents

Part III   A Social, Ethical and Legal Analysis of Human ICT Implants

Contributors

Bernhard Anrig is a Professor of Computer Science and director of studies at the Division of Computer Science, Bern University of Applied Sciences, Switzerland. He studied Mathematics and Computer Science at the University of Fribourg, Switzerland where he completed his Ph.D. on "Probabilistic Model-Based Diagnostics". He worked as a Senior Researcher at the University of Fribourg, Switzerland on reasoning under uncertainty, especially probabilistic argumentation systems and information systems, information algebras as well as the connections between reliability analysis and diagnostics. He is member of the Research Institute for Security in the Information Society at the Bern University of Applied Sciences (www.risis.ch) where his current domain of research and activities cover security and privacy, identities, virtual identities, anonymisation, and web-service security.

Diana M Bowman is an Assistant Professor in the Risk Science Centre and the Department of Health Management and Policy, School of Public Health, the University of Michigan. With a background in both science and law (BSc, LLB, Ph.D.), Diana's research has primarily focused on legal, regulatory, and public health dimensions of new technologies such as nanotechnologies. She has also published in the area of injury prevention, public health law, and intellectual property rights. Diana is the co-editor of *Nanotechnology Environmental Health and Safety: Risk, Regulation and Management* (with Hull, 2010) and *The International Handbook on Regulating Nanotechnologies* (with Hodge and Maynard, 2010). In 2010, Diana took up a position as a member of the Australian Government's National Enabling Technology Strategy's Expert Forum. She is admitted to practice as a Barrister and Solicitor of the Supreme Court of Victoria (Australia).

Ramón Compañó is the Program Manager of the JRC-IPTS (Institute for Prospective Technological Studies, Joint Research Centre, European Commission). Ramón graduated in Physics from the Technical University (RWTH) of Aachen, and completed his Ph.D. at the RWTH Aachen (Germany) and the University of

Modena (Italy). He holds a Masters degree in Technology Management from Solvay/ULB (Brussels). In 1993, he joined the European Commission taking care of R&D actions in support of less favored regions, and small and medium enterprises in the fields of standardisation, advanced materials, physical analysis, and metrology. In 1996, he contributed to the R&D strategy and implementation of the EC programmes on nanoelectronics domain being responsible for the evaluation and follow-up of R&D projects. Later, in the Strategy for ICT Research & Development Unit at DG Information Society, he contributed in updating the work programme. In 2004, he joined the Institute for Prospective Technological Studies, where he analysed the impacts of emerging ICTs on society and economy. Since 2010 he is the Program Manager of the JRC-IPTS overseeing the scientific strategy of the institute.

Barbara Daskala is employed in the European Network and Information Security Agency (ENISA), where she currently works on promoting information security and privacy risk management practices and particularly in identifying emerging security and privacy risks posed by new technologies and ways to address them. Before that, she worked in the Institute for Prospective Technological Studies (IPTS) of European Commission's Joint Research Centre, where she was involved in research on the social implications of emerging and future technologies and in particular, of the ambient intelligence environment. Prior to that, she had been working for about 4 years in Ernst & Young in Athens, Greece, as an IT auditor and information security consultant. She is a Certified Information Systems Security Professional (CISSP)® and a Certified Information Systems Auditor (CISA)®. She also holds an MSc in Analysis, Design, and Management of Information Systems from the London School of Economics & Political Science, UK and a BSc in Business Administration from the University of Piraeus, Greece.

Claude Fuhrer is a Professor of Computer Science at the Bern University of Applied Sciences in Bienne (Switzerland). He graduated in Microtechnology from the Bern University of Applied Science and in Computer Science from the faculty of sciences of the University of Neuchâtel. His main research interest are the development of mobile applications, the physical aspects of image rendering and geometrical computations applied to computer graphics. Claude Fuhrer teaches introductory courses in java and C programing as well as specialisation courses in computer graphics (splines and NURBS).

Mark N. Gasson was a Senior Research Fellow at the School of Systems Engineering, University of Reading in the UK and now holds the post of Visiting Research Fellow. He obtained his first degree in Cybernetics and Control Engineering in 1998 from the Department of Cybernetics at Reading, and obtained his Ph.D. in 2005 for developing an experimental invasive interface between the nervous system of a human volunteer and a computer system. His interdisciplinary research interests predominantly centre around user-centric applications of emerging technologies, with specific focus on pushing the envelope of Human–Machine interaction. In 2009, he demonstrated the privacy issues of creating detailed behavioral profiles using GPS in mobile phones, and the following year became the first human to be infected by a computer virus using an RFID device implanted in his hand. In 2010 he was the General Chair for the IEEE International Symposium on Technology and Society (ISTAS 2010) in Australia. Mark is passionate about getting young people engaged in and excited about science and aims to generate interest by making the topics accessible and relevant through a variety of activities. He has delivered invited public lectures and workshops internationally and has participated in over 100 events in the last 10 years. In order to communicate in the widest arena possible, he also engages with the national press and other media outlets, including as a guest expert for BBC news.

Mireille Hildebrandt is Professor of Smart Environments, Data Protection and the Rule of Law at the Institute of Computing and Information Sciences (ICIS), Radboud University Nijmegen; Associate Professor of Jurisprudence at the Erasmus School of Law (ELS) in Rotterdam and Senior Researcher at Law Science Technology and Society (LSTS), Vrije Universiteit, Brussels. After obtaining her law degree and a Ph.D. in legal philosophy she has been involved in various research projects on profiling, identification technologies, and Ambient Intelligence. She teaches legal philosophy to lawyers and 'law in cyberspace' to computer scientists, and publishes on the nexus of law, philosophy and smart environments. In 2008, she co-edited *Profiling the European Citizens. Cross-Disciplinary Perspectives* (Springer), together with Serge Gutwirth and in 2011 she co-edited *Law, Human Agency and Autonomic Computing. The Philosophy of Law meets the Philosophy of Technology* (Routledge), together with Antoinette Rouvroy. She is co-editor of Criminal Law and Philosophy, and of the Netherlands Journal of Legal Philosophy. See http://works.bepress.commireille_hildebrandt/.

Bert-Jaap Koops is a Professor of Regulation & Technology at the Tilburg Institute for Law, Technology, and Society (TILT) of Tilburg University, the Netherlands. From 2005–2010 he was a member of De Jonge Akademie a young-researcher branch of the Royal Netherlands Academy of Arts and Sciences. His main research interests are law & technology, in particular criminal-law issues in investigation powers, privacy, computer crime, identity-related crime, DNA forensics. He is also interested in other topics of technology regulation, including privacy data protection, identity, digital constitutional rights, 'code as law', human enhancement. He has co-edited six books in English on ICT regulation, including *Starting Points for ICT Regulation* (2006), *Cybercrime and Jurisdiction* (2006), *Constitutional Rights and New Technologies* (2008), and *Dimensions of Technology Regulation* (2010) he has published many articles and books in English and Dutch on a wide variety of topics.

Eleni Kosta is a Senior Legal Researcher at the Interdisciplinary Centre for Law and ICT (ICRI) of the Katholieke Universiteit Leuven in Belgium. Eleni joined ICRI in the summer of 2005, where she conducts research in the field of privacy and identity management, specialising in electronic communications and new technologies. She worked on various European funded research projects, such as

PRIME and PICOS and was involved in the FIDIS Network of Excellence and in the Thematic Network Privacy OS. She is currently working on the European FP7 research projects + Spaces (Policy simulation in virtual worlds) and EXPERI-MEDIA (Experiments in live social and networked media experiences). Before joining ICRI, Eleni obtained her Law degree at the University of Athens in 2002 and in 2004, she obtained a Masters degree in Public Law at the same University. In the academic year 2004–2005, she followed the LL.M. Program in Legal Informatics (Rechtsinformatik) at the University of Hanover (EULISP) with a scholarship from the Greek State Scholarships Foundation (IKY). In June 2011, Eleni obtained the title of Doctor of Law with a thesis entitled "Unravelling consent in European data protection legislation—a prospective study on consent in electronic communications" under the supervision of Prof. Dr. Jos Dumortier. She is also a member of the Heraklion BAR Association and a part-time associate at the Brussels based law firm time.lex.

Ronald Leenes is a Professor in Regulation by Technology at TILT, the Tilburg Institute for Law, Technology, and Society (Tilburg University). His primary research interest is regulation of, and by, technology, in particular applied to privacy and identity management. Leenes received his Ph.D. for a study on hard cases in law and Artificial Intelligence and Law from the University of Twente. Ronald was work package leader in the EU FP6 PRIME project, the FP7 PrimeLife project, and the EU FP7 project ENDORSE. Currently, he leads the TILT team in the EU Robolaw project which investigates the way in which emerging technologies in the field of (bio-) robotics (e.g. bionics, neural interfaces and nanotechnologies) have a bearing on the content, meaning and setting of the law. He has contributed to and edited various deliverables for the EU FP6 Network of Excellence 'Future of IDentity in the Information Society' (FIDIS). Ronald is chair of IFIP WG 9.6/11.7 IT misuse and the law. He regularly serves on the program commission of various international conferences.

Arnold Roosendaal LLM MPhil studied Dutch Law and obtained an LLM in Law and Technology. After his LLM, he followed a Research Masters Programme, for which he obtained his MPhil. Currently Arnold is a Ph.D. candidate at the Tilburg Institute for Law, Technology and Society (TILT), Tilburg University. He is also a partner at Fennell Roosendaal Research and Advice. He has a great interest in law and technology and the implications of technological developments on society. In his research he specifically looks at the implications for individuals, often by analysing effects on privacy and autonomy of the individual. Arnold has participated in several international research projects, such as FIDIS and Prime-Life, and has written several international publications. Next to that, he regularly participates in conferences as a speaker or panalist.

Bibi van den Berg has an MA (with honours) and a Ph.D. in philosophy, both obtained from Erasmus University Rotterdam in the Netherlands. She specialises in philosophy of technology and philosophical anthropology. After completing a Ph.D. Bibi had a postdoc position at the Tilburg Institute for Law, Technology and

Society (TILT) of Tilburg University in the Netherlands. Her research focused on social, legal, and ethical issues in the deployment of autonomous technologies and robots, and on the relationship between human enhancement and robotics. She also conducted research into questions relating to identity and privacy in online worlds. Currently, Bibi holds a position as an Assistant Professor at eLaw, the Centre for Law in the Information Society at Leiden University's Law School. At eLaw, Bibi conducts research into many ways in which regulators could use technologies to influence behaviour of citizens, and into questions of legitimacy and democracy in relation to this kind of technological influence.

Paweł Rotter received a Ph.D. degree from the Faculty of Electrical Engineering, AGH University of Science and Technology in Kraków where he currently holds the post of Assistant Professor in the Automatics Department. In the past, he has worked at the Artificial Intelligence and Data Analysis Laboratory of the Aristotle University of Thessaloniki (2001–2002), at the Cracow University of Technology (2002–2004), and at the Institute for Prospective Technological Studies of the EC Joint Research Centre in Seville (2005–2008). He was engaged in 12 research projects and is the author or co-author of over 25 publications and many technical reports, chapters in world-range monographs and a series of popular science articles. His main research interests are computer vision and multicriteria optimisation.

Ryszard Tadeusiewicz obtained his Master of Science degree with honours at the AGH University of Science and Technology in 1971. Additionally, after receiving his degree in Automatic Control Engineering, he studied at the Faculty of Medicine at the Medical Academy in Krakow, as well as undertaking studies in the field of mathematical and computer methods in economics. He has written over 600 scientific papers, which have been published in prestigious Polish and international scientific journals, as well as numerous conference articles. Professor Tadeusiewicz has also written over 70 scientific monographs and books, among them are highly popular textbooks, which have had many editions. He was supervisor of 56 and reviewer of more than 200 doctoral theses. In 2007, Polish scientists elected him to be the President of IEEE Computational Intelligence Society—Polish Chapter.

Abbreviations

AmI	Ambient intelligence
ATM	Automatic teller machine
BCI	Brain–computer interface
CAS	Court of Arbitration for Sport
DBS	Deep brain stimulation
EEG	Electroencephalogram
EGE	European Group on Ethics
EMGs	Electromyograms
EU	European Union
FDA	Food and drug administration
fMRI	Functional magnetic resonance imaging
HCI	Human–computer interface
ICD	Implanted cardiac defibrillator
ICT	Information and communication technologies
ID	Identification
IMDs	Implantable medical devices
KDD	Knowledge discovery in databases
LBS	Location-based services
LFPs	Local field potentials
MEMS	Micro electro-mechanical systems
NRA	National Rifles Association
OECD	Organisation for Economic Co-operation and Development
PD	Parkinson's disease
PIN	Personal identification number
RF	Radio frequency
RFID	Radio frequency idenification
STN	Sub-thalamic nucleus
TETs	Transparency enhancing tools

UDHR	Universal Declaration on Human Rights
UK	United Kingdom
US/USA	United States of America
VIP	Very important person

Chapter 1
Human ICT Implants: From Invasive to Pervasive

Mark N. Gasson, Eleni Kosta and Diana M. Bowman

Abstract While considered by many to be within the realm of science fiction, for several decades information and communication technology (ICT) has been implanted into the human body. Advanced medical devices such as cochlear implants and deep brain stimulators are commonplace and research into new ways to invasively interface with the human body are opening up new application areas such as retinal implants and sensate prosthetics. It is apparent that as these implantable medical technologies continue to advance their potential for human enhancement, i.e. enabling abilities over and above those which humans normally possess, will become increasingly attractive. In the first instance, this may be giving a person with a deficient sense a device that enables them to function on a vastly superior level. However, it is clear that healthy people will look to implantable technology to augment what we would consider their 'normal' abilities. Technology enthusiasts have already begun to realise the potential of simple

M. N. Gasson (✉)
Visiting Research Fellow, School of Systems Engineering,
University of Reading, Berkshire, UK
e-mail: m.n.gasson@reading.ac.uk

E. Kosta
Faculty of Law, Senior Legal Researcher in the Interdisciplinary Centre
for Law & ICT (ICRI), KU Leuven, Belgium
e-mail: eleni.kosta@law.kuleuven.be

D. M. Bowman
Department of Health Management and Policy and the Risk Science Centre,
School of Public Health, The University of Michigan,
Ann Arbor, MI, USA
e-mail: dibowman@umich.edu

M. N. Gasson et al. (eds.), *Human ICT Implants: Technical, Legal and Ethical Considerations*,
Information Technology and Law Series 23, DOI: 10.1007/978-90-6704-870-5_1,
© T.M.C. ASSER PRESS, The Hague, The Netherlands, and the author(s) 2012

1

implantable technologies, with people opting to have passive silicon devices surgically implanted to facilitate identification. It is equally foreseeable that the application of implantable technology, developed initially in a medical context, will be repurposed to augment the abilities of healthy humans. Such developments are beginning to redefine our relationship with technology. The changes are not just technological—they are driving changes in cultural and social paradigms, and further empowering people to seek new experiences and employ new services. It is evident that we need to address the incipient technical, legal, ethical and social issues that the development of human ICT implant devices may bring. This chapter gives an overview of the landscape of issues surrounding human ICT implants, and explains how the following chapters in this book serve to address these key areas in more depth.

Is the human body a suitable place for a microchip? Such questions, and the discussions that they provoke, are no longer hypothetical. Indeed, they have not been for some time now. While considered by many to be within the realm of science fiction, human implants which incorporate information and communications technologies (ICT) have been developed in a medical context for decades and routinely implanted into people.

Medical devices such as cardiac defibrillators and pacemakers used to restore heart rhythm and cochlear implants to restore hearing have become well established and are widely used throughout the world as a way in which to improve an individual's well-being and public health more generally.

These sophisticated devices form intimate links between technology and the body. As of 2009, about 188,000 people worldwide have received cochlear implants, and promising trials have been conducted with retinal and sub-retinal implants for vision. These devices are designed to (partially) repair deaf and blind people's impairments, allowing them to (re)gain 'normal' sensory perception. Indeed the application of implantable technology for medical use is typically 'restorative', i.e. it aims to restore some deficient ability. Recent developments in engineering technologies have meant that the ability to integrate silicon with biology is reaching new levels and medical devices that interact directly with the brain are becoming commonplace. A prominent example is Deep Brain Stimulator (DBS) technology for the treatment of Parkinson's disease.

Keeping in step with developments of other fundamental technologies, these types of devices are becoming increasingly complex and capable, with their peripheral functionality also continuing to grow. Data logging and wireless, real-time communications with external computing devices are now well within their capabilities and are becoming standard features, albeit without due attention to inherent security and privacy implications. In Chap. 4, Tadeusizicz et al. explore the state-of-the-art of invasively implantable technologies and show how cutting edge research is feeding into devices being developed in a medical context. The authors focus their analysis on four technologies—pacemakers and cardiac defibrillators, cochlear implant technology, DBS and brain computer interfaces for sight restoration.

It is apparent that, as implantable medical technologies continue to advance, their potential for human enhancement, i.e. enabling abilities over and above those which humans normally possess, will become increasingly attractive. In Chap. 2, Gasson describes and discusses such enhancements. In the first instance, this may be giving a person with a deficient sense a device that enables them to function on a vastly superior level. Consider, for example, a retinal implant that provides the user with telescopic and night vision and can receive and relay information. However, it is equally foreseeable that healthy people may look to the technology to augment what we would consider their 'normal' abilities (i.e. typical functioning for an individual) through implantable technology.[1] This may well hold moral significance that needs consideration.[2] Experiments demonstrating human enhancement through the implantation of technology in healthy humans have been performed for over a decade by some academic research groups. Indeed, some such widely publicised experiments to surgically implant technologies have already taken steps forward by linking the human nervous system directly to a computer.

More recently, technology enthusiasts have begun to realise the potential of simple implantable technologies such as glass capsule Radio Frequency IDentification (RFID) transponders. This new trend for low-tech human ICT implants has recently risen in the public consciousness with people opting to have such passive silicon devices surgically implanted to allow identification and tracking, which is discussed in detail by Rotter et al. in Chap. 3.

In their early application, these implantable devices had very simple functionality—the ability to broadcast a fixed unique identifier over a short range on request. While largely deployed for animal identification, the implantable tags commercialised for human use have the same function—an identifier that could be cross-referenced with a database that holds all other information. However, the core technology has continued to develop. Although non-implantable RFID devices in general remain more advanced than implantable glass capsule types, these continue to evolve as well, which opens up new possibilities, and new issues, as articulated in Chap. 2. It has been argued that implantable RFID devices have now developed to the point whereby we should consider the devices themselves as simple computers. While this has begun to introduce challenging questions, the radically improved capability over previous generations of the technology has been vividly demonstrated by the infection with a computer virus of an RFID device implanted in a human. Although the numbers of people with such devices are still small, the commercialisation of this technology means that they could become commonplace. The security and privacy implications of RFID for a

---

[1] Daniels 2000.

[2] Lin and Allhoff 2008.

variety of applications have been well explored, but the use of them inside the body serves to further aggravate some of the known issues.

It is evident that advances in technology, coupled with a change of perception and attitudes within society, partly driven by increasing familiarity, are enabling new opportunities for human enhancement. Such enhancement includes any activity by which we improve our bodies, minds or abilities—things we do to enhance our well being,[3] of which implantable devices is only one part. The willingness of self-experimenters to push the boundaries is of great importance, but it is the gradual evolution of technology which enables the most notable change. While developments in the area of human ICT implants have been slow to come, the convergence of fields such as nanotechnology, biotechnology, information technology, cognitive science, robotics and artificial intelligence is likely to increase the application and prevalence of such human ICT implant devices. It is considered that the next wave of disruptive technologies will actually be a result of this domain fusion rather than from any one field in isolation.

As such, it cannot be assumed that since the technology is immature, there is no need to address the incipient legal, ethical and social issues that the development of human ICT implant devices may bring. Current applications of such implants alone introduce challenging questions. Indeed the increasing commercialisation has generated debate over aspects of the technology and its products.

Questions over potential risks and liability are discussed in detail in Chaps. 5–7. In Chap. 5, Rotter et al. focus on potential privacy risks associated with the increasing deployment of RFID technology in humans. Their analysis does not, however, conclude at that point, with the authors also considering potential medical risks associated with implants. Rather than wait for such risks to be realised, the authors argue that controls and measures can be developed and engineered now in order to minimise such risks. This analysis is then built upon in Chap. 6 by Rotter and Gasson in regards to medical devices in particular. As with Chap. 5, the authors do not merely identify the issues, but rather seek to provide some answers as to how such concerns may be addressed by practitioners in their various fields.

In Chap. 7, Roosendaal provides an analysis on the debate relating to medical liability and product liability in the context of human ICT implants. The debate on the distinction between therapy and enhancement is briefly touched upon and the importance of causation is discussed. Challenging issues specific to human ICT implants, such as security of the ICT components, are also introduced. The chapter focuses on the fact that the complexity of the implants, as well as the myriad of people involved in their development, programming and placing lead to possible difficulties in determining who can be held liable for damages and the extent of information duties of manufacturers and healthcare practitioners.

Hildebrandt and Anrig, in Chap. 11, focus on a variety of ethical implications of human ICT implants. The authors explain how different ethical implications arise

---

[3] Allhoff et al. 2011.

from different types of implants, depending on the context in which they are used. After a first assessment of what is at stake, the chapter briefly discusses the Opinion 20 of the European Group on the Ethics of Science and New Technologies on the ethical aspects of ICT implants in the human body. The discussion is extended further by tracing the ethical implications for democracy and the rule of law, considering the use of implants for the repair as well as the enhancement of human capabilities.

Developments in implantable technologies for enhancement are beginning to redefine our relationship with technology. The changes are not just technological— they are driving changes in cultural and social paradigms, and further empowering people to seek new experiences and employ new services. The basic foundations of advanced implant devices are being developed for clear medical purposes and it is reasonable to assume that few would argue against this progress for such noble, therapeutic causes. Equally, cosmetic surgery has demonstrated that we cannot assume people will refuse a procedure because it is highly invasive. So, while we may still be some way away, there is clear evidence that implantable devices capable of significant human enhancement will become reality, and most probably applied in applications beyond their original purpose. Technological advancement is a part of our evolution, and the significant next step of forming direct bi-directional links with the human brain is moving inexorably closer. This will certainly open up the potential for many new application areas. Scientists have indicated for some time that a human/machine symbiosis—a physical linking of the two entities such that humans can seamlessly harness the power of machine intelligence and technological capability—is a real possibility. The typical inter- face through which a user currently interacts with technology provides a distinct layer of separation between what the user wants the machine to do and what it actually does, which imposes a considerable cognitive load. The main issue is interfacing the human motor and sensory channels with the technology in a reliable, durable, effective, bi-directional way.

One possible solution is to avoid this sensorimotor bottleneck altogether by interfacing directly with the human nervous system. Indeed some predict that within the next 30 years neural interfaces will be designed in a way that will not only increase the dynamic range of senses, but will also enhance memory, enable "cyberthink"—invisible communication with others and technology[4] and increase creativity and other abstract facets of the human mind. However, this could evidently give a bionically enhanced person a competitive advantage over the nonenhanced. Koops and Leenes, in their chapter—Chap. 10—discuss the normative implications of this *hypothetical* form of human enhancement, focusing on aspects that are particularly relevant to this type of enhancement as compared to existing and well-discussed other forms of enhancement. In particular, the chapter examines information asymmetries, ethical aspects related to human enhancement and some legal issues, where the information advantage of bionic sensory implants

---

[4] McGee and Maguire 2007.

could make a difference. Based on this discussion, Koops and Leenes highlight a range of questions that deserve further reflection and provide suggestions for the regulatory response to address the challenges posed by the future of bionic sensory implants.

While the technology possibilities are becoming clearer, some commentators have begun to question whether the option to have an ICT human implant is—or should be—a right of the individual. Many argue that advantages gained by a person who has an implant probably imply a relative disadvantage for those who do not. The concern is that if the advantages are great enough, perhaps even having implications on employment or other significant aspects of life, then this may create a two-tier society, the haves and the have-nots. Ultimately this would lead to increased pressure on people to have an implant. If indeed they can.

Other commentators have argued that because of technology, such a divide already exists and no distinction should be made between carrying an external device and having the device implanted in the body. While it is evident that such technologies will require some sort of regulation, it is clear that strong regulation or a ban on development will never be globally implemented, leaving the advantages to be harnessed by the more liberal parts of the world.[5] Evidently a balance must be struck between stringent regulation and individual liberty.[6] However, from a wider legal perspective, human ICT implants could have other major impacts on the people carrying them. Indeed there may be negative consequences, resulting in damages. To enable these damages to be compensated it has to be clear who can be held liable. Is it the healthcare practitioner who placed the implant, the manufacturer, the person who programmed or updated the software, or does the user carry all the risks themself?

The use of human ICT implants also raises many wider questions: for example, if an implant interferes or alters our deliberative process, then are we acting freely while under the influence of the enhancement?[7] However, human ICT implants may further have various implications for human rights. In particular, the right to bodily integrity may be at stake, due to the direct connection to the human body. Several twentieth-century philosophers have pointed out that when humans use artefacts and technologies, these often tend to become extensions of their bodies: through extended physiological proprioception they become incorporated into the user's body schema. Most of us can flawlessly park a car or write with a pen because of this principle. Technologies, although 'other', can become 'part' of a user's bodily repertoire, even if they are not embedded into the human body.

At the same time, it is interesting to note that in some cases technologies can be experienced as 'alien', or that they can even lead users to feel 'alienated' from themselves. The former may happen when we are new at using a technology, or when it malfunctions or breaks down. The latter has been shown to occur, for example,

---

[5] Hughes 2004.
[6] Greely 2005.
[7] Guston et al. 2007.

in patients who undergo DBS. After treatment, these patients sometimes state that they feel estranged from themselves, that they no longer feel they are the same person. Van den Berg, in Chap. 12, uses some of the central ideas from philosophy of technology to clarify these two (seemingly contradictory) perspectives.

In Chap. 8, the scope of the right to bodily integrity is discussed by Roosendaal. A more detailed discussion on how the right is affected by implants is largely based on biotechnological implants. In this field, the distinction between therapy and enhancement, as well as implications of implants for the concept of the body and its integrity has been largely debated. Subsequently, the right to bodily integrity is approached with a focus on other, non-living, implants. Specific emphasis is placed on nanotechnology-based implants and information carriers and some additional challenges for human rights, other than bodily integrity, resulting from ICT implants are described.

As already discussed above, the increasing commercialisation of human ICT implants has generated heated debate over the ethical, legal and social implications of their use. Despite stakeholders calling for greater policy and legal certainty within these areas, gaps have already begun to emerge between the commercial reality and the current legal frameworks designed to regulate it. In Chap. 9, Kosta and Bowman, focus on a special issue and examine the effectiveness of the European Union current data protection regulatory framework for regulating ICT implants. By examining current and future applications of human ICT implants, the research presented within Chap. 9 highlights the potential regulatory challenges posed by the applications, and makes a series of recommendations as to how such issues may be best avoided by jurisdictions grappling with similar emerging issues. In doing so, the chapter draws together the notions of innovation, risk and data protection within the context of a broader governance framework.

Having seen the applications which RFID has found as an implantable technology despite earlier pessimism, we evidently need to consider the application of the more advanced medically orientated technologies on healthy individuals, i.e. enhancement rather than restoration, as a distinct probability. While it is necessary to acknowledge that our continuing evolution may well mean that we all become part machine, we must also be mindful of the new threats this step brings. It is clear that a number of issues stem from such human enhancement and it is timely to have debates to address the wider implications. Thus, clear consideration needs to be given now to the fundamental moral, ethical, social, psychological and legal ramifications of such enhancement technologies, especially as these issues typically lag far behind technology development.

While it is impossible to know whether the utopian, dystopian or pedestrian predictions of human ICT implants are realisable in the long term, the debate has an immediate bearing on the real world with conclusions that could affect researchers, manufacturers, social institutions as well as our ideals of freedom and human dignity.[8] However, it has taken the wider academic community some time to agree

---

[8] Lin 2007.

that meaningful discourse on the topic of human implantable ICT technology is of value. Indeed, the term 'cyborg' (a blend of cybernetic and organism) was until very recently largely met with derision, yet this is evidently a very real scenario. As developments in implantable medical technologies point to greater possibilities for human enhancement, this shift in thinking is not too soon in coming.

References

Allhoff F, Lin P, Steinberg J (2011) Ethics of human enhancement: an executive summary. Sci Eng Ethics 17:201–212

Daniels N (2000) Normal functioning and the treatment-enhancement distinction. Camb Q Healthc Ethics 9:309–322

Greely H (2005) Regulating human biological enhancements: Questionable justifications and international complications. The mind, the body, and the law: University of Technology, Sydney, Law Review 7:87–110. Santa Clara J Int Law 4:87–110 (2006) (joint issue)

Guston D, Parsi J, Tosi J (2007) Anticipating the ethical and political challenges of human nanotechnologies. In: Allhoff F, Lin P, Moor J, Weckert J (eds) Nanoethics: the ethical and social implications of nanotechnology. Wiley, Hoboken

Hughes J (2004) Citizen cyborg: why democratic societies must respond to the redesigned human of the future. Westview Press, Cambridge

Lin P (2007) Nanotechnology bound: evaluating the case for more regulation. NanoEthics 2: 105–122

Lin P, Allhoff F (2008) Untangling the debate: the ethics of human enhancement. NanoEthics 2:251–264

McGee EM, Maguire GQ (2007) Becoming borg to become immortal: regulating brain implant technologies. Camb Q Healthc Ethics 16(3):291–302

Part I
Human ICT Implants:
From Restoration to Enhancement

Chapter 2
Human ICT Implants: From Restorative Application to Human Enhancement

Mark N. Gasson

Abstract Human ICT implants such as cochlear implants and cardiac pacemakers have been in common clinical use for many years, forming intimate links between technology and the body. Such medical devices have become increasingly advanced in their functionality, with some able to modify behaviour by directly interacting with the human brain and others coming closer to restoring functionality which outperforms its natural counterpart. More recently, and somewhat more controversially, low-tech human ICT implants have been increasingly employed in healthy people, in non-therapeutic contexts. Applications typically focus on identification such as VIP entry into nightclubs, automated payments and controlling access to secure facilities. While reviewing the state of the art, this chapter makes the case that with the desire of technology enthusiasts and self-experimenters to push boundaries, increasing familiarity driving cultural and societal changes, advances in medical technology and the inevitable drift of medical technology to non-medical application, this is clearly just the beginning for human enhancement using ICT implants.

Contents

M. N. Gasson (✉)
Visiting Research Fellow, School of Systems Engineering,
University of Reading, Berkshire, UK
e-mail: m.n.gasson@reading.ac.uk

M. N. Gasson et al. (eds.), *Human ICT Implants: Technical, Legal and Ethical Considerations*,
Information Technology and Law Series 23, DOI: 10.1007/978-90-6704-870-5_2,
© T.M.C. ASSER PRESS, The Hague, The Netherlands, and the author(s) 2012

2.1 Introduction

There is a fair range of 'restorative' devices already in clinical use, although many, such as artificial joints, could not through their function alone be considered ICT devices. Others, such as the artificial heart pacemaker, have become notably sophisticated in recent years with integrated sensors to adjust performance based on estimated demand, internal logging of biological data, and RF communication with the outside world.

These battery powered devices are known as Implantable Medical Devices (IMDs), and have been used for decades as lifesaving devices, becoming increasingly capable and multifunctional. The operation of such devices is not necessarily based on a routine that is periodically repeated—some actions are performed automatically as a result of a continuous monitoring of the patient's body by sensors embedded in the IMD. With such devices, there is usually a need for two-way communication between IMDs and the external world.

Data can be sent to the implanted device in order to programme its parameters or to trigger a specific action on demand. The device may in turn send data, for example the history of any measured characteristics or a log of device actions, which can be externally analysed. The most effective way of such communication is through wireless, radio transmission. However, such communication can raise concerns about privacy and security.

The main communication functions of IMDs are (see Fig. 2.1 below)

- Communicating the presence and type of device to medical staff
- Deactivation of the device on demand, e.g. before an operation
- Sending collected measured medical data (e.g. heart rate and body temperature) and data on the history of the device operation to help in the diagnostic process[1] and the auditing of the device's operational history
- Configuration of the device
- Manual control over the device
- Upgrade of the software included in the device

Of great interest is the development of IMD technologies that are able to interact with us on a neural level. The most ubiquitous sensory neural prosthesis is by far the cochlear implant[2]; where destruction of cochlea hair cells and the related degeneration of auditory nerve fibres has resulted in sensorineural hearing loss, the prostheses is designed to elicit patterns of nerve activity via a linear array of electrodes implanted in the deaf patient's cochlea that mimics those of a normal ear for a range of frequencies, (as discussed in detail by Tadeusiewicz et al. in Chap. 4). Current devices enable around 20% of those implanted to communicate without lip reading and the vast majority to communicate fluently when the sound

---

[1] Some modern IMDs offer the possibility of online home monitoring: data received wirelessly from the IMD by the base station are passed through a website to a doctor.

[2] Zeng 2004, p 1.

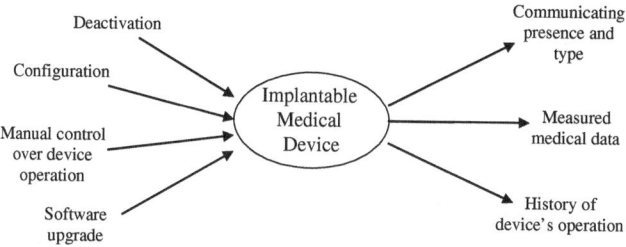

Fig. 2.1 Communication of a generic IMD with the external world, showing the directional flow of information

is combined with lip reading. Its modest success is related to the ratio of stimulation channels to active sensor channels in a fully functional ear, with recent devices having up to 24 channels, while the human ear utilises upwards of 30,000 fibres on the auditory nerve. With the limitations of the cochlear implant in mind, the artificial visual prosthesis[3] is certainly substantially more ambitious. While degenerative processes, such as retinitis pigmentosa, selectively affect the photo-detectors of the retina, the fibres of the optic nerve remain functional, so with direct stimulation of the nerve it has been possible for the recipient to perceive simple shapes and letters. However, the difficulties with restoring full sight are several orders of magnitude greater than those of the cochlear implant simply because the retina contains millions of photodetectors that need to be artificially replicated, and so this technology remains in development. Sensate prosthetics is another growing application area of neural interface technology, whereby a measure of sensation is restored using signals from small tactile transducers distributed within an artificial limb. The transducer output can be employed to stimulate the sensory nerve fibres remaining in the residual limb that are naturally associated with a sensation. This more closely replicates stimuli in the original sensory modality, rather than forming a type of feedback using neural pathways not normally associated with the information being fed back. As a result, it is supposed that the user can employ lower level reflexes that exist within the central nervous system, making control of the prosthesis more subconscious.

While cochlear implants, retina stimulators and sensate prosthetics operate by artificially manipulating the peripheral nervous system, less research has been conducted on direct electrical interaction with the human central nervous system, and in particular the brain.[4] A Brain-Computer Interface (BCI) consists of hardware and software for direct communication between the brain and external devices. Information may be passed from the brain to an external device, for example to enable control of a computer or robot arm with thought or in the opposite direction

---

[3] Hossain et al. 2005, p 30.

[4] The following part of this subsection contains contributions from Ryszard Tadeusiewicz and Pawel Rotter.

such as from an artificial eye to the brain. The performance achievable highly depends on the degree of invasiveness:

- In a non-invasive BCI, no implant is used. The most common non-invasive method is based on recordings of brain activity taken from the outside surface of the head, known as an electroencephalogram (EEG). Other methods include magnetoencephalography and magnetic resonance imaging. Non-invasive BCIs typically pass information only in one direction (rarely to the brain). Signals are attenuated by the skull so non-invasive BCIs cannot achieve a performance comparable with those based on implanted electrodes.[5] On the other hand, they do not require an operation, do not cause side effects, and are easy to put on and to take off.
- An invasive BCI includes electrodes implanted directly into the grey matter of the brain, which requires a complicated and risky surgical procedure. There are a number of health and ethical concerns associated with this approach. Implanted electrodes enable applications which are unachievable by non-invasive methods.[6] However, most invasive BCIs monitor multi-neuronal intracortical action potentials, requiring an interface which includes sufficient processing in order to relate recorded neural signals with e.g. movement intent. The need to position electrodes as close as possible to the source of signals, and the need for long-term reliability and stability in a hostile environment are common issues.
- In partially invasive BCI electrodes are implanted into the skull but outside of the brain. The operation is easier and brings less medical risks than in the case of invasive BCI, but the expected functional performance is lower.

The selection of the method to be employed—invasive or non-invasive—will depend on the application. In some cases invasive methods must be employed because of the complexity of the application. However, non-invasive methods can be used for simple tasks in a range of fields, from entertainment to medical applications.

Simple control of external devices by intentionally modifying brain activity is possible using basic, non-invasive methods. In experiments carried out in mid-1990s with paralysed patients, intended variations of slow cortical potentials in their EEG were changed into binary signals so they could step-by-step select letters and write messages.[7] A further line of research has centred on invasive implants for patients who have suffered a stroke resulting in paralysis. An early example is the use of a '3rd generation' brain implant which enables a physically incapable brainstem stroke victim to control the movement of a cursor on a computer screen.[8] Functional Magnetic Resonance Imaging (fMRI) of the

---

[5] Popescu et al. 2008, p 78.

[6] Graimann et al. 2011.

[7] Winters 2003.

[8] Kennedy et al. 2000, p 198 and Kennedy et al. 2004, p 72.

subject's brain was initially carried out to localise where activity was most pronounced whilst the subject was thinking about various movements. A hollow glass electrode cone containing two gold wires and a neurotrophic compound (giving it the title 'Neurotrophic Electrode') was then implanted into the motor cortex, in the area of maximum activity. The neurotrophic compound encouraged nerve tissue to grow into the glass cone such that when the patient thought about moving his hand, the subsequent activity was detected by the electrode, then amplified and transmitted by a radio link to a computer where the signals were translated into control signals to bring about movement of the cursor. With two electrodes in place, the subject successfully learnt to move the cursor around by thinking about different movements. Eventually the patient reached a level of control where no abstraction was needed—to move the cursor he simply thought about moving the cursor. Notably, during the period that the implant was in place, no rejection of the implant was observed; indeed the neurons growing into the electrode allowed for stable long-term recordings.

The year 2005 saw the first surgery being carried out which enabled a quadriplegic patient to control their artificial hand using an invasive implant.[9] However, hand precision control is still generally unsatisfactory with these devices, even after long and frustrating training. It is therefore questionable whether currently the benefits justify substantial health risks (including infection and implant rejection), plus the economic costs associated with surgery. At this early stage, there is no wide deployment of this technology, although research has demonstrated its potential. Christian Kandlbauer was the first man with a mind-controlled arm prosthesis who attained his driving licence. His story is an example of success of this technology but his tragic death after an accident in a car he was driving highlights the technical, legal and ethical questions surrounding these developments.

Work on animals,[10,11] has demonstrated how direct brain stimulation can be used to guide rats through a maze problem, essentially by reinforcement, by evoking stimuli to the cortical whisker areas to suggest the presence of an object, and stimulation of the medial forebrain bundle (thought to be responsible for both the sense of motivation and the sense of reward) when the rat moves accordingly. Early work to translate this research to humans demonstrated radical (and occasionally dubiously interpreted) changes in mood and personalities when such 'pleasure centres' were stimulated[12,13]. This period saw some 70 patients implanted with permanent micro-stimulators to treat a variety of disorders with reportedly good success, although the indiscriminate use of the procedure and significant failure rate saw it largely condemned. This may have been in part because the disorders targeted were psychiatric rather than neurological, and it was

---

[9] Hochberg 2006, p 164.

[10] Olds and Milner 1954, p 419.

[11] Talwar et al. 2002, p 37.

[12] Moan and Heath 1972, p 23.

[13] Delgado 1977, p 88.

not until the 1980s, when French scientists discovered that the symptoms of Parkinson's disease (PD), with better understood anatomical pathology, were treatable using Deep Brain Stimulation (DBS), that research again picked up pace (see Chap. 4 of this book). However, difficulties in accurately targeting structures deep in the brain, lack of safe durable electrodes, problems of miniaturising electronics and power supply limitations meant that such therapy was not readily available for several more years.

The ability of electrical neural stimulation to drive behaviour and modify brain function without the recipient's cognitive intervention is evident from this type of device. Further, it has been demonstrated how electrical stimulation can be used to replace the natural percept, for example the work by Romo et al.[14] However, in all cases these devices operate in a unidirectional fashion—the ability to form direct bi-directional links with the human nervous system certainly opens up the potential for many new application areas. Nevertheless, bi-directional neural implants are very much experimental. Whilst they have much potential in the areas of prosthetics, major developments have been slow in coming. Recent research in the area of DBS has shown that by recording brain activity via the implanted electrodes used for DBS in Parkinson's patients it is possible to detect characteristic signal changes in the target nuclei prior to the event of tremor, and so stimulation based on a prediction of what the brain will do is possible.[15] The development of such technologies, which are able to decode the brain's function, are clearly of great value.

2.2 Application of ICT Implants for Enhancement

The application of human implants is largely medical, and the vast majority of these devices are not ICT devices. However, such passive devices have been utilised for many years for enhancement of healthy people through varying degrees of body modification from simple cosmetic improvements, body decoration to radical reshaping of the physical form. In 2006, self-experimenters began to utilise small sub-dermal neodymium magnet implants, typically under the skin of the fingertips, for sensory experimentation whereby the movement of the implant in the presence of magnetic fields can be felt by the individual. More recently this type of implant has been used to convert non-human sensory information, such as sonar or distance, into touch information by manipulating the implant via external electromagnets to control stimulation of sensory receptors.[16] However, largely because of the increased technical complexities, healthy people have only started to explore how implantable ICT technology can be harnessed to enhance their normal abilities.

---

[14] Romo et al. 2000, p 273.

[15] Gasson et al. 2005a, p 83.

[16] Hameed et al. 2010, p 106.

Radio Frequency IDentification (RFID) technology was originally developed for automatic identification of physical objects. An RFID tag—a small device attached to the object—emits identification data through radio waves in response to a query by an RFID reader. This information is captured by the reader and then further processed. RFID technology has been increasingly employed as a 'barcode replacement' device due to the number of advantages that it offers. The RFID devices have been increasingly used in production lines and the logistics chain of enterprises and are starting to penetrate other sectors including, for example, medical and healthcare, defence and agriculture.

Governments around the world and industry itself have been keen promoters of RFID technology. All European Union (EU) Member States, the United States of America, Australia and many other countries are gradually deploying electronic passports. These new passports contain RFID tags that store personal data, including biometric data from the passport holder. This allows for semi-automatic authentication of people at borders. Credit-card-sized contactless smart cards are another example of an application based on RFID technology.[17] Such smart cards are becoming increasingly popular for access control. While some RFID-enhanced smart cards contain only identification numbers, other cards include additional cryptographic security features to protect the data during transmission. Sophisticated RFID-based devices not only identify, they can also be used to track people's location and activities.[18]

Whilst RFID tags have been commonly used for uncontroversial applications such as the supervision of stock and other animals for some time now, humans are increasingly coming to the fore. RFID implants that are introduced into the human body have already been commercialised; such implants were specifically designed to facilitate the identification and authentication process. RFID implants for identification and authentication of people provide some potential advantages compared to other established methods. The identification process is, for example, fully automatic and convenient: no typing or confirming of information, no remembering of password or carrying a token. Moreover, there is no need neither for the person to clean their hand(s) before putting it under the fingerprint scanner, nor having to stand still as they would have to do if they were having an iris scan done. Identification and authentication with an RFID implant is practically immediate; there is no loss of time associated with, for example, the typing of a password, for acquiring and matching of biometrics or for taking a smart card out of a wallet.

It can be argued that implants are also an extremely reliable method of identification, especially compared to biometrics, which—due to the statistical nature of their matching process—cannot guarantee error-free results.[19] While implants

---

[17] For example see Mifare, http://www.mifare.net.

[18] Smith et al. 2005, p 39.

[19] See for more detail FIDIS deliverable D6.1 'Forensic Implications of Identity Management Systems' and D3.10 'Biometrics in identity management'.

may require replacement during a person's lifetime, they are considered more durable than tokens and many types of biometrics, which usually change due to ageing. Unlike tokens, implants cannot be lost or stolen (unless an attacker extracts the implant). RFID implants can be used by everyone without exception, including people with cognitive impairment. The user will always be identifiable, even if they are unconscious or not carrying any identity documents.

Commercial RFID implants for people are best described as 'passive tags'—i.e. they do not require built-in batteries but operate by making use of the energy emitted by the external RFID reader. As a result, and since they have no moving parts, once implanted under the skin they can be operational for more than a decade. Some manufacturers have even made the claim that they will be operation for more than 40 years.

Their extremely small size and lack of any internal power source does however limit the devices' performance in terms of memory, processing power and communication range. These hardware limitations make it difficult to design RFID implants with advanced authentication methods. The limited communication range, is not necessary a drawback; it may also be an advantage from a security and privacy point of view.

The first recorded human RFID implantation occurred on Monday, 24 August 1998, when a groundbreaking experiment was conducted by Prof. Warwick's group at the University of Reading in the UK. At the heart of this work were the sub-dermal implantation of an RFID tag and the augmentation of the infrastructure at the university's Department of Cybernetics with RF nodes such that the system was able to track him, via the tag, as he roamed the building. The possibilities using this technology were, even at that time, not greatly limited, however the system was restricted to simple profiling of his behaviour. From this, automated customisation of his environment was possible, such as unlocking doors, turning on lights and brewing his coffee on arrival.

While the public response to this work was varied, from suggestions that this was the work of the devil,[20] to awe of the technological possibilities, acknowledgement of the prophetic merit largely mirrored that of academic musings on the scientific value. Few could appreciate the idea that people may actually be open to having such devices implanted if there was some net benefit in doing so. Equally, few entertained the realisation that, at that time, RFID technology was on the cusp of becoming cost effective enough to essentially become ubiquitous.

Some 6 years later, implantable identifying RFID tags were commercialised by 'VeriChip' and approved by the FDA in the USA for human use (for more discussions regarding the VeriChip, see Chap. 3 of this book). It was proposed that

---

[20] Revelation 13:16–18 "He [the beast] also forced everyone, small and great, rich and poor, free and slave, to receive a mark on his right hand or on his forehead, so that no one could buy or sell unless he had the mark, which is the name of the beast or the number of his name". Such scaremongering is in keeping with the flawed logic which demonstrates that the common barcode contains a hidden '666', e.g. as described by Relfe in her 1982 book "The New Money System: 666".

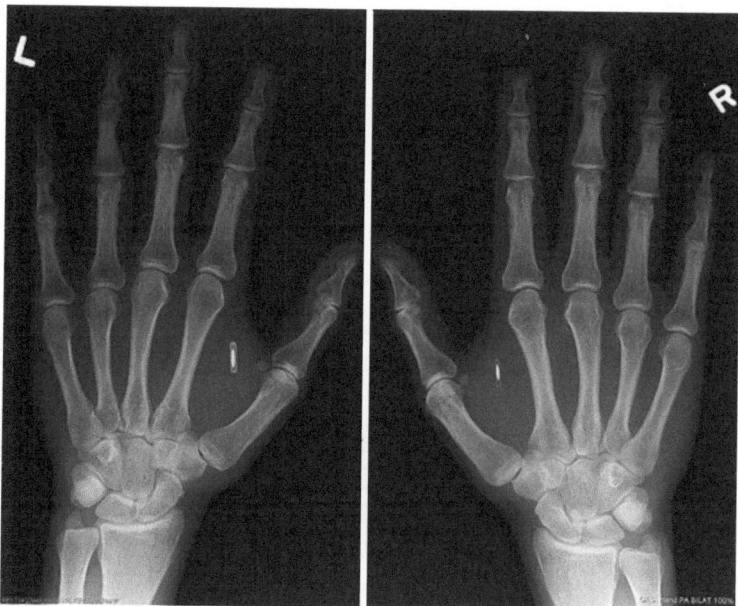

Fig. 2.2 An individual with two RFID implants: His left hand contains a 3 by 13 mm EM4102 glass RFID tag that was implanted by a cosmetic surgeon, his right hand contains a 2 by 12 mm Philips HITAG 2048S tag with crypto-security features, implanted by a GP using an animal injector kit (Graafstra 2007, p 18.)

these devices could essentially replace 'medic alert' bracelets and be used to relay medical details when linked with an online medical database. Such devices have subsequently been used to allow access to secure areas in building complexes, for example the Mexican Attorney General's office implanted 18 of its staff members in 2004 to control access to a secure data room, and nightclubs in Barcelona, Spain and The Netherlands use a similar implantable chip to allow entry to their VIP customers, and enable automated payments. By 2007, reports of people implanting themselves with commercially available RFID tags for a variety of applications became a familiar occurrence (see, for example, Fig. 2.2). The broad discussion on security and privacy issues regarding mass RFID deployment has since picked up pace, and security experts are now specifically warning of the inherent risks associated with using RFID for the authentication of people.[21]

Whilst the idea that RFID can be used to covertly track an individual 24-7 betrays a fundamental misunderstanding of the limitations of the technology, there

---

[21] RFID technology is still in its infancy and resource-constraints in both power and computational capabilities make it hard to apply well-understood privacy protection techniques that normally rely heavily on cryptography. For instance, a 'man-in-the-middle' attack would make it possible for an attacker to steal the identity of a person (i.e. tag identifier), while widely published techniques for RFID tag cloning make utilising this information technically feasible.

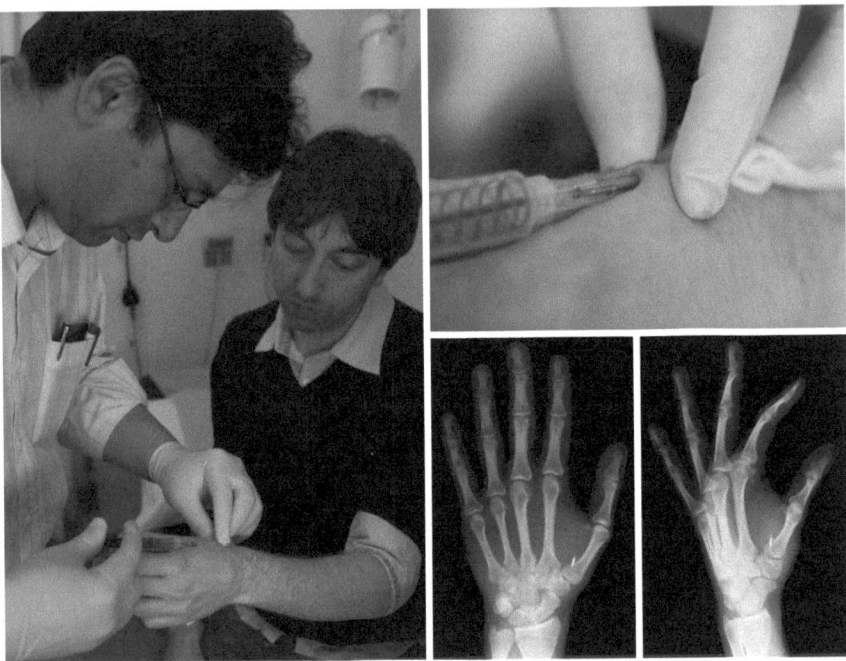

Fig. 2.3 An RFID tag is injected into the left hand of the author by a surgeon (*left*), shown in close up (*top right*). Two X-ray images taken post-procedure (*bottom right*) show the position of the tag in the hand near the thumb

are genuine concerns to address. The use of implanted RFID tags in this scenario is especially thwart with issues because being implanted forms a clear, permanent link with the individual and makes compromised tags hard to revoke.

In the early applications as an implantable device, RFID tags had very simple functionality—the ability to broadcast a fixed unique identifier over a short range on request. While largely deployed for animal identification, the implantable tags commercialised for human use had the same function—an identifier which could be cross-referenced with a database that held all other information. However, the core technology has continued to develop, and although non-implantable RFID devices in general remain more advanced than implantable, glass capsule types, these too continue to evolve which opens up new possibilities, and new issues. To further explore this, the earliest experiments with an implanted RFID device conducted in 1998 were revisited after ten years using the latest in implantable RFID technology.

On 16 March 2009, the author had a glass capsule HITAG S 2048 RFID device implanted into his left hand, as illustrated by Fig. 2.3.[22] While containing a 32-bit unique identifier number, similarly to older devices, the device also has a 2048-bit

---

[22] Gasson 2010, p 61.

read/writable memory to store data and the option of 48-bit secret key based encryption for secure data transfer. These are clear advances over the older implantable technology, which could only broadcast a fixed identifier, and enable new applications to be realised. As in the 1998 study, the tag was used as an identification device for the University of Reading's intelligent building infrastructure. A mobile phone was also augmented with a reader such that only the user with the correct tag could use the phone. In the 1998 study, simple profiles were constructed of users of the building, based on tracking their movements and preferences, which were stored on a central database.

Because data can be stored on the latest generation of implantable tags, in a modification, this profile information was stored both in the building's database and on the implanted HITAG S tag such that the user could enter a new building, which could then access the profile data. Updates to the profile were generated centrally, and written to the tag if it needed updating. While this is seemingly a useful extension to the original system, it comes coupled with new threats.

In 2006, researchers from Vrije Universiteit in Amsterdam demonstrated how commercially available RFID tags could be used to spread malicious computer code.[23] In order to do this the devices required the ability to store data and interact with a potentially vulnerable database system. To demonstrate the concept, a large form factor RFID sticky label tag was infected with a piece of malicious code and used to contaminate a database. However, despite the provocative paper title practically all implantable RFID devices at that time, typically being only readable or of very low data storage capacity, were not actually vulnerable to this. However, four years later as a proof of concept, the implanted HITAG S tag was successfully infected with malicious code containing a computer virus.[24] Because of the way the malicious code had been written, instead of simply reading data from the implanted tag to store in the database, the system also executed some SQL injection code, overwriting in the database valid data by a copy of the virus in such a way that any tag subsequently using the system will likely become overwritten and infected. A feature of a computer virus is that is must have the ability to self-replicate, and this is evident here. Having corrupted the database contents in such a way to allow replication, there is a further 'payload' (some additional malicious activity) associated with the virus. Administration of the database is typically done through a web browser, and once the system is infected the web browser is redirected to another website, denying easy access to rectify the problem. More potentially harmful payloads have previously been demonstrated[25] including enabling unauthorised system access.

Concerns for those who have decided to have an RFID tag implanted are valid, although an assumption is that such procedures will never become compulsory and so most people will remain unaffected. However, while mass deployment of RFID

---

[23] Rieback et al. 2006, p 169.

[24] Gasson 2010, p 61.

[25] Rieback et al. 2006, p 169.

technologies is well documented, especially in the context of commerce, it should be noted that, through non-nefarious means, it is possible that people could become implanted with RFID unknowingly. This is mostly related to safety issues regarding passive medical devices, such as hip replacements and breast implants, whereby being able to determine the exact manufacturing details non-invasively could be advantageous. This is especially valuable when manufacturing faults are subsequently discovered and devices of unknown provenance have been used. Thus, embedding an RFID enabled device in a unit before it is surgically utilised would enable this function. Further, following the polemic on silicone-gel breast implants[26] which resurfaced with a vengeance in early 2012 following a global health scare triggered by unapproved materials being used in Poly Implant Prothese (PIP) breast implants,[27] a device based around RFID technology, designed to be located inside the breast, which detects rupture has been developed, and many are investigating the benefits of being able to non-invasively monitor the condition of a medical device, such as a heart valve, using this type of technology. However, all these applications result in the wider issue of having RFID implanted.

Exact numbers of those who have received this type of low-tech implantable technology are not known, but it is clear that the figure is rising, and, with familiarity, public acceptance will surely grow. Because such uses of the technology were largely dismissed as improbable from the outset, a lack of timely debate on the wider implications means that we are now faced with the prospect of addressing them whilst the technology gets a foothold. Not least of all, this certainly leaves some open questions which technologists must now address. It is not hard to imagine that dealing with technical and wider issues retrospectively will be immensely more difficult. The potential application areas for implantable RFID are further explored in Chaps. 3 and 4.

2.3 Where Restorative Meets Enhancement

The relatively new trend for having passive RFID implants has recently risen in the public consciousness, although less publicised developments of high-tech implants in the medical domain have been progressing for several decades. Indeed, a significant drive behind the development of implantable devices is medical—i.e. restoring deficient abilities in humans. Given this, there are two clear routes by which technology developed for restorative application may ultimately lead to enhancement. The first is that it is conceivable that a piece of technology designed as a restorative device may actually give the recipient a capability which exceeds the normal human ability it is designed to replace. For example, advances in cochlear implants may result in the recipient having vastly improved hearing over

---

[26] Kessler 1992, p 1713.
[27] http://www.bbc.co.uk/news/health-16391522. Accessed 05 January 2012.

that of a normal human that could then be considered enhancement. The discussion in this context has begun on the topic of prosthetic limbs.[28] There are, however, no clear examples relating to implantable technology to date, although Moore[29] describes the case of a patient with an artificial heart who found he could use the device to lower his heart rate to help falling asleep.

The second is the application of implantable technology, developed initially in a medical context, to augment the abilities of healthy humans. Reports of this pioneering step are rare, although in a notable echo of 1998, the University of Reading in the UK has been active in this area. On March 14th, 2002, an array of 100 individual needle electrodes was surgically implanted into the median nerve fibres of the left arm of Professor Kevin Warwick, a healthy volunteer,[30,31] (see also[32] for a personal account). This study demonstrated, in a rudimentary fashion, a range of applications, from nervous system to nervous system communication, feedback control of robotic devices and augmented sensory capabilities.

To date, there are no studies involving implantation in the central nervous system of healthy volunteers that have been well reported. There is, however, some largely anecdotal evidence of the occasional positive side effect that DBS has had in patients. In one such case, a graphic designer, who received DBS surgery for a severe Tourettes disorder, found that stimulation through one specific electrode could actually make her more creative. Indeed, when this electrode was used, her employer noted an improvement in colour and layout in her graphic design work.[33] The application of this type of effect in the long term clearly cannot be discounted, and so nor can the translation of medical devices to enhancement. Indeed, the ability to form direct, bi-directional links with the human brain will open up the potential for many new application areas. Scientists have indicated for some time that a human/machine symbiosis—a physical linking of the two entities such that humans can seamlessly harness the power of machine intelligence and technological capability—is a real possibility. The typical interface through which a user currently interacts with technology provides a distinct layer of separation between what the user wants the machine to do and what it actually does, which imposes a considerable cognitive load. The main issue is interfacing the human motor and sensory channels with the technology in a reliable, durable, effective, bi-directional way. One possible solution is to avoid this sensorimotor bottleneck altogether by interfacing directly with the human nervous system. While still in its infancy, scientists predict that within the next

---

[28] Camporesi 2008, p 639.

[29] Moore 2008.

[30] Gasson et al. 2005b, p 365.

[31] Warwick et al. 2003, p 1369.

[32] Warwick 2002.

[33] Cosgrove 2004.

30 years neural interfaces will be designed that will not only increase the dynamic range of senses, but will also enhance memory, enable "cyberthink"—invisible communication with others and technology[34] and increase creativity and other abstract facets of the human mind.

2.3.1 Human Enhancement and Bodily Boundaries

While being a clear demonstration of how implantable devices are becoming more complex, capable and potentially vulnerable,[35] being susceptible to malicious attack, e.g. a computer virus, also raises interesting questions linked to the concept of the body.

As functions of the body are restored or further enhanced by implanted devices, the boundaries of the body become increasingly unclear. Previous recipients of RFID implants echo the sentiments of many cochlear implant and heart pacemaker users—the implant becomes perceived as being part of the body.[36] That is, what the user understands to be their body includes the technological enhancement. In essence, the boundaries between man and machine simply become theoretical. This development in our traditional notion of what constitutes our body and its boundaries leads to two notable repercussions here. Firstly, it becomes possible to talk in terms of a human (albeit a technologically enhanced human) becoming e.g. infected by a computer virus. Thus, in that light, the simple experiment conducted by the author's group in 2009 gave rise to the world's first human to be infected by a computer virus. Secondly, this development of our concept of the body impacts on certain human rights, in particular the right to bodily integrity. Bodily integrity constitutes a right to do with one's body whatever one wants (a right to self-determination) and it implies the right to prevent one's body from being harmed by others. In this context, a computer virus infecting an implanted device constitutes an infringement on the right to bodily integrity. These issues are further considered by Roosendaal in Chap. 8. A number of wider moral, ethical and legal issues stem from applications of these technologies,[37,38,39] and it is difficult to foresee the social consequences of adoption long term which may fundamentally change our very conception of self and sense of identity. It is clearly timely to have further and rigorous debate regarding the use of implantable technology in individuals for human enhancement.

---

[34] McGee and Maguire 2007, p 291.

[35] Maisel and Kohno 2010, p 1164.

[36] Warwick 2003, p 131.

[37] Rodotà and Capurro 2005, p 18.

[38] Hansson 2005, p 519.

[39] Kosta and Gasson 2008.

2.4 Terminology and Crucial Distinctions[40]

For a proper understanding of the wide scope of human implants we will briefly discuss the different types of implants and their different functions. Technological implants are often used to restore bodily functions, as in the case of cardiac pacemakers or deep brain stimulation (see Chap. 4). Moreover, entirely different uses can be envisioned, like monitoring of biological functions to enable real time diagnostics or the identification of clients or patients in order to grant them a right of access or to streamline the information system of healthcare institutions. The most imaginative usage of human implants is human enhancement or the creation of human–machine hybrids that challenge our notion of what it is to be human and raise the issue of who is in control: software programmes, the individual human mind of whoever 'has' or 'is' the implant, or the data controller who holds the remote control.

Due to the different affordances of distinct types of implants in the human body, and the different consequences they may have, we will discriminate between:

1. Implants that aim to *restore* or *repair* human capabilities
2. Implants designed for the *diagnosis* of a biological state
3. Implants that *identify* a person in order to e.g. provide access to certain locations, information or knowledge, or to automatically pay/bill for services rendered
4. Implants that aim to *enhance* human memory, vision, auditory perception, alertness or other human capabilities

Technically, we will discriminate between:

1. *Active* implantable devices which can function using an internal power source, and *Passive* implantable devices which depend on power supplied to it remotely
2. *Online* ICT implants that rely for their operation on an online connection to an external computer and *Offline* or stand alone ICT implants that can operate independently of external devices

2.5 Conclusion

For many years it has been all too easy to dismiss the idea of implanting technology in our bodies to enhance our abilities as science fiction and simply improbable. To many, violating their bodies in such a way is unthinkable regardless of the (albeit seemingly unrealistic) benefits it may bring. While the layperson holding this position is to some degree understandable, surprisingly it

---

[40] This section was written by Mireille Hildebrandt.

has also taken the wider academic community some time to agree that meaningful discourse on the topic of human implantable ICT technology is of value. Indeed, the term 'cyborg' (a blend of cybernetic and organism) was until very recently largely met with derision. This chapter has sought to highlight in what ways this scenario is evidently a very real possibility.

Advances in medical technologies are notoriously slow in coming, not necessarily because the enabling technology does not exist, but largely due to the safeguards surrounding their efficacy, safety and commercialisation. Equally, there is a substantial gap in our combined knowledge regarding how the human body, and in particular the brain, actually functions on a low level. As research and new instrumentation, such as medical imaging technology, allows us unparalleled access to the fundamental workings of the body, and gives hints as to how the abstract components which form our complex personalities manifest; we will find new ways to apply technology to manipulate them. Already familiar implantable medical technology exists which is designed to interact with the human body on an intimate level. The capabilities of cardiac pacemakers, which include wireless communication, far exceed most people's expectations, and deep brain stimulators, which are implanted in their thousands globally every year, interact directly with the brain. The use of these devices is tantamount to having a computer system implanted in the body and this, and the issues it brings, is described further in Chaps. 4 and 6 using a sample of state-of-the-art medical devices. It is clear, especially as new medical applications are found, that many of us will evidently end up with some sort of implanted piece of computing technology at some point in our lives.

Developments in implantable medical technologies also point to greater possibilities for human enhancement. Here there are two distinct routes by which this may happen. Restorative technology aims to repair or ameliorate some form of deficient functionality, with most implantable medical devices falling into this category. However, the device could well give the user functionality which out performs the 'normal' range, or has additional functional elements. Consider a retinal implant which gives someone their sight back, but in a form that is twice as efficient as the human eye, and with an augmented display for additional information. This is a form of human enhancement. Interesting questions then arise as to whether a person with 'normal' vision should then be able to 'upgrade' themselves using this technology. Indeed the desire for people with no medical need looking to harness the opportunities presented by medical technologies is the second route by which human enhancement can be envisaged. If, for example, deep brain stimulator technology can be repurposed to give better memory, more creativity, a different sense, or a new form of communication, then surely a market for this will appear. It may well be that the benefits of whatever is achievable, and that we can only begin to imagine, will mean that in reality there is no real option to having it. There are obvious parallels here to the wide uptake of technologies such as mobile phones, computers and use of the internet.

The foundation for the novel deployment of technologies grounded in medical devices has been set by the willingness of self-experimenters to push the boundaries. The more advanced and invasive examples of these have largely been in the academic research domain, but the recent phenomenon of even basic RFID implants for a variety of applications is also a hugely important milestone in this evolution. It is clearly impossible to discount the use of invasive, implantable technologies for human enhancement, and acknowledgement of this as fact is of utmost importance as we consider how we may need to deal with the host of changes and challenges it will bring.

References

Camporesi S (2008) Oscar pistorius, enhancement and post-humans. J Med Ethics 34:639

Cosgrove GR (2004) Neuroscience, brain, and behavior V: Deep brain stimulation. Transcript—session 6, June 25, 2004, The President's Council on BioEthics. http://bioethics.georgetown.edu/pcbe/transcripts/june04/session6.html. Accessed 15 Apr 2011

Delgado JM (1977) Instrumentation, working hypotheses, and clinical aspects of neurostimulation. Appl Neurophysiol 40(2–4):88–110

Gasson MN (2010) Human enhancement: Could you become infected with a computer virus? IEEE international symposium on technology and society, ISTAS 2010. Wollongong, Australia, pp 61–68

Gasson MN, Hutt BD, Goodhew I, Kyberd P, Warwick K (2005a) Invasive neural prosthesis for neural signal detection and nerve stimulation. Int J Adapt Control Signal Process 19(5): 365–375

Gasson MN, Wang SY, Aziz TZ, Stein JF, Warwick K (2005b) Towards a demand driven deep-brain stimulator for the treatment of movement disorders. MASP2005, 3rd IEE international seminar on medical applications of signal processing, London, UK, 3–4 Nov 2005, pp 83–86

Graafstra A (2007) Hands On. IEEE Spectr 44(3):18–23

Graimann B, Allison B, Pfurtscheller G (2011) Brain-computer interfaces non-invasive and invasive technologies. Springer, New York

Hameed J, Harrison I, Gasson MN, Warwick K (2010) A novel human-machine interface using subdermal magnetic implants. Proceedings IEEE international conference on cybernetic intelligent systems, Reading, pp 106–110

Hansson SO (2005) Implant ethics. J Med Ethics 31(9):519–525

Hochberg LR (2006) Neuronal ensemble control of prosthetic devices by a human with tetraplegia. Nature 442:164–171

Hossain P, Seetho IW, Browning AC, Amoaku WM (2005) Artificial means for restoring vision. BMJ 330:30–33

Kennedy P, Andreasen D, Ehirim P, King B, Kirby T, Mao H, Moore M (2004) Using human extra-cortical local field potentials to control a switch. J Neural Eng 1(2):72–77

Kennedy P, Bakay R, Moore M, Adams K, Goldwaith J (2000) Direct control of a computer from the human central nervous system. IEEE Trans Rehabil Eng 8:198–202

Kessler DA (1992) The basis of the FDAs decision on breast implants. N Engl J Med 326(25):1713–1715

Kosta E, Gasson MN (eds) (2008) A study on ICT implants. FIDIS. http://www.fidis.net. Accessed 15 Apr 2011

Maisel WH, Kohno T (2010) Improving the security and privacy of implantable medical devices. New Engl J Med 362(13):1164–1166

McGee EM, Maguire GQ (2007) Becoming borg to become immortal: regulating brain implant technologies. Camb Q Healthc Ethics 16(3):291–302

Moan CE, Heath RG (1972) Septal stimulation for the initiation of heterosexual activity in a
    homosexual male. J Behav Ther Exp Psychiatry 3:23–30
Moore P (2008) Enhancing me: the hope and the hype of human enhancement. Wiley, New York
Olds J, Milner PM (1954) Positive reinforcement produced by electrical stimulation of septal area
    and other regions of rat brain. J Comp Physiol Psychol 47:419–427
Popescu F, Blankertz B, Müller KR (2008) Computational challenges for noninvasive brain
    computer interfaces. IEEE Intell Syst 23:78–79
Rieback M, Crispo B, Tanenbaum A (2006) Is your cat infected with a computer virus?
    Proceedings 4th annual IEEE international conference on pervasive computing and
    communication (PERCOM06), Pisa, Italy, 13–17 March, pp 169–179
Rodotà S, Capurro R (eds) (2005) Ethical aspects of ICT implants in the human body. Opinion of
    the European group on ethics in science and new technologies to the european commission,
    pp 18–23D
Romo R, Hernandez A, Zainos A, Brody CD, Lemus L (2000) Sensing without touching:
    psychophysical performance based on cortical microstimulation. Neuron 26:273–278
Smith JR, Fishkin K, Jiang B, Mamishev A, Philipose M, Rea A, Roy S, Sundara-Rajan L (2005)
    RFID-based techniques for human-activity detection. Commun ACM (Special issue on RFID)
    48:39–44
Talwar SK, Xu S, Hawley ES, Weiss SA, Moxon KA, Chapin JK (2002) Rat navigation guided
    by remote control. Nature 417:37–38
Warwick K (2002) I, Cyborg, Century
Warwick K (2003) Cyborg morals, cyborg values, cyborg ethics. Ethics Inf Technol 5:131–137
Warwick K, Gasson MN, Hutt BD, Goodhew I, Kyberd P, Andrews BJ, Teddy P, Shad A (2003)
    The application of implant technology in Cybernetic systems. Arch Neurol 60(5):1369–1373
Winters J (2003) Communicating by brain waves psychology today. http://www.psychology
    today.com/articles/200307/communicating-brain-waves. Accessed 15 Apr 2011
Zeng FG (2004) Trends in cochlear implants. Trends Amplif 8(1):1–34

Chapter 3
Potential Application Areas for RFID Implants

Pawel Rotter, Barbara Daskala, Ramon Compañó, Bernhard Anrig and Claude Fuhrer

Abstract Radio Frequency IDentification (RFID) technology was originally developed for automatic identification of physical objects. An RFID tag—a small device attached to the object—emits identification data through radio waves in response to a query by an RFID reader which also supplies it power. RFID technology has been increasingly employed as a 'barcode replacement' due to the number of advantages that it offers and has been used in production lines and the logistics chain of enterprises and are starting to penetrate other sectors including medical and health care, defence and agriculture. While the first recorded human implantation of an RFID device was in 1998, in 2004, the first RFID implant device was approved for human use by the United States Food and Drug

P. Rotter (✉)
AGH University of Science and Technology,
Krakow, Poland
e-mail: rotter@agh.edu.pl

B. Daskala
European Network and Information Security Agency (ENISA), Heraklion, Greece
e-mail: Barbara.Daskala@enisa.europa.eu

R. Compañó
European Commission Joint Research Center, Institute for Prospective Technological Studies (IPTS), Seville, Spain
e-mail: Ramon.Compano@ec.europa.eu

B. Anrig
Division of Computer Science, Bern University of Applied Sciences, Berne, Switzerland
e-mail: bernhard.anrig@bfh.ch

C. Fuhrer
Division of Computer Science, Bern University of Applied Sciences, Bienne, Switzerland
e-mail: claude.fuhrer@bfh.ch

M. N. Gasson et al. (eds.), *Human ICT Implants: Technical, Legal and Ethical Considerations*, Information Technology and Law Series 23, DOI: 10.1007/978-90-6704-870-5_3,
© T.M.C. ASSER PRESS, The Hague, The Netherlands, and the author(s) 2012

Administration. No data about the owner *per se* is stored on the device, instead the ID number points to a corresponding entry in a centralised database and can be used to facilitate identification and authentication. The continued commercialisation of RFID implant devices approved for human use, along with a trend for technology enthusiasts and self-experimenters to implant a variety of more advanced RFID technology points the way to future application of these devices. Further explored in this chapter is the use of human implantable RFID in the areas of patient identification in health care, access to services, as a complementary tool for other identification methods, access control for mobile devices, smart environments and other potential longer term applications.

Contents

3.1 Patient Identification in Health care

In October 2004 the first RFID implant for human use—the so-called VeriChip—was approved for human use (as a medical device) by the United States (US) Food and Drug Administration (FDA). The VeriChip implant is a small passive (RFID) tag which stores a unique identification (ID) number. It can typically be read from a distance of up to 10–15 cm.[1] The ID number contains enough digits so as to potentially uniquely identify *everybody* in the world. No data about the owner *per se* is stored on the device, instead the ID number points to a corresponding entry in a centralised database.

The first commercial application—aptly named the VeriMed—was designed for the health care sector. An authorised user including, for example, a doctor, could access a patient's medical data on the VeriMed database (note that the VeriMed database is unlikely to comprise a patient's entire medical record) through a

---

[1] The VeriChip has been designed to operate at a distance of about 10 cm with a handheld reader and 50 cm with a door reader but cannot operate over very large distances. Simulations show that a standard distance can be increased several times (up to 0.5 m for ISO 14443), but with further increase the signal disappears in the environment noise.

password-protected website; this was done by using the patient's implant ID number, detected by an RFID reader, as the database key.

Application drift of this solution is likely to occur over the coming decades within the healthcare context. Persons suffering from diseases such as, for example, epilepsy or Alzheimer's, may wish to be implanted with such a device so that they can be tracked within their environment. In such instances it would be possible to detect if the person had left their 'safe' environment, determine if medication had been taken, or if a person had fallen. While such monitoring can already be successfully performed through applications that are external to the body, an implantable device is superior to an external device as it cannot be removed (without significant skill) or lost.

On 7 August 2006, the first known life-saving incident involving a VeriChip was reported. Policeman W. Koretsky, a VeriChip subscriber, was rushed to a medical centre with a head trauma following a high-speed police pursuit which resulted in a crash. Doctors were able to access his medical records in the VeriMed database, using the ID number retrieved from his VeriChip implant, and tender the requisite care needed to preserve his life in the time critical situation.

3.2 Access to Services

An example of how the technology may be employed in a non-health related field was provided by a nightclub in Barcelona, Spain, in 2004. The club, the Baja Beach Club, invited their 'most valued' customers to be implanted with an RFID device. Using a syringe type delivery system, the tag was implanted on site by a trained employee of the Baja Beach Club. The implant not only granted the customer access to the VIP areas, but also—and arguably more importantly— enabled the selected clientele to use the device as a payment method. Waiters were equipped with a scanner which registered the identity of the customer and charged her account (e.g. a credit card) with the cost of the drinks ordered.

While the implant device has now been used for over 5 years, there appears to have been little or no expansion of the system's core purpose so far or wide adoption of this application elsewhere. The possible misuse of information through 'silent' capture and processing of the tags (for example, a person's location) merits hesitation. Using the Baja Beach Club example, one may infer with enough precision the habits of a customer and in principle an individual or company could establish a user profile containing information about, for example but not limited to[2]:

- the frequency of visits
- the time/s of the year that the customer is in the region (location)
- the usual time when the customer visits the night club (beginning of the evening or end of the night) and the duration of these visits (even if the customers are not scanned when they go, using the time of their last order, one can estimate when they left)

---

[2] Note that the points hereafter are speculation on what could be done with the data.

- the consumption habits, including the quantity and type of drink combined with the consumption frequency
- based on the number of drinks the customer has paid within some time slot (when they pay for a round of drinks, for example), one could estimate the size of their party. A customer who regularly invites many people may be considered a 'prescriber', may have a higher influence on other people, and is of interest to the club. On the other hand, a customer who is always alone may be of less interest, and
- considering the events organised by the Club at the time the customer was present, one could deduce the kind of activities that the customer liked (musical preferences, the shows she likes, etc.). This parameter is even more important if the customer is a 'prescriber'. This could be used to determine the kind of advertisements she should receive.

If one goes some steps further, one could imagine that a venue, such as the Baja Beach Club, could be augmented with several scanners which could then be used to characterise the person based on their movements and their proximity to other tagged individuals.

The automatic identification enables people to skip queues and (probably most importantly) makes them feel 'special', 'important' and/or 'distinguished'. In such applications, the driving motivation behind the implantation of the RFID device is their convenience (immediate identification, difficult to lose or break). These advantages, however, should be balanced with the ethical, privacy and health concerns that the use of RFID implants entail. In simple terms: is it worth being 'chipped' for not queuing for a drink?

3.3 RFID Implants as a Complementary Tool for Other Identification Methods

RFID implants in combination with established identification and authentication technologies can provide additional security and may be used as the second or third factor in a multimodal identification and authentication. Such implants could be used to protect systems from accepting PINs/passwords obtained by theft or fraud, or which have been cracked or disclosed without authorisation. Similarly, the RFID implants could be employed to reduce the risk of false authentication with a stolen token or with one which has been loaned by an authorised person to a third party, either willingly or not. Prominent applications are those where people are less willing to delegate their rights, such as withdrawing cash at an automatic teller machine.

In closed environments—such as restricted workplaces—the identity of employees may be verified at the entrance to the property or building; a lower level of security may then be employed within the area. Such a framework may be used to prevent authorised people from supplying their credentials to unauthorised users. For instance, employees' movements from one room to another could be tracked by placing RFID readers in all doorways. While the primary motivation for doing so would be for security purposes, a secondary benefit is also possible: in the case of an accident, for example, the location of every worker would be known.

RFID implants may also be used in secure, but more open environments, such as a hospital. The device could be used to assist in the prevention of unauthorised access to restricted wards, rooms or secure cupboards. In such environments RFID implants could also be used to facilitate the identification of a person. Where strong authentication is required, RFID implants can be complemented with other identification and authentication technologies, like PINs/passwords, tokens and/or biometrics.

Reading an identification number from an RFID implant may also speed-up the process of biometric identification. To identify a person, their biometrics must be compared with each sample in a database. The RFID implant would speed-up the process by providing an immediate and reliable identification (1:N), while biometrics would subsequently offer a strong authentication (comparing the user's sample with only one database entry, indicated by the RFID implant).

The next generation of RFID implants may incorporate some biometric information[3] so that the user and implant are able to authenticate each other. Such a system would be secure against a coercive attack as an implant extracted from a victim's body could not be used by another person (who has different biometric features). A drawback of any such system is the possibility that an attacker is able to duplicate the identification number and modify biometric features.

Finally, it should be noted that many RFID devices, including the commercialised VeriChip device, are susceptible to cloning—i.e. the duplication of the 'unique' identifier in another device.[4] As such, it has been proposed that implanted RFID tags are only used for identification and not authentication. Despite this, in 2004, several employees in the organised-crime division of the Mexican Attorney General's offices received implants giving them access to restricted areas (see the case study at the end of this chapter for more details). In 2006, an Ohio-based company had tags implanted into some of its employees and CityWatcher.com, a private video surveillance company, uses the technology for controlling access to a room where it holds security video footage for government agencies and the police.

3.4 Access Control for Mobile Devices

Mobile devices, such as notebook computers, PDAs or mobile phones have the capacity to store a growing amount of confidential information about their owners, many of who are motivated to secure them. In existing devices, people

---

[3] Perakslis and Wolk 2006, p 34.
[4] Halamka et al. 2006, p 699.

are usually identified when they switch on the device. When a user leaves, the device still retains the authentication for a certain time period during which an unauthorised person may gain access to the device. RFID implants have the potential to reduce or minimise the chance of unauthorised access in such instances. An RFID-based system can detect continuously the presence of the authorised person and demand a re-authentication when this person leaves the area covered by the reader. Thus, the additional security feature arises from the permanent identification through the proximity of the RFID implant. For high-security applications, the mobile device would identify a person by reading the RFID implant, and then the owner would authenticate to the device via a PIN/password or biometrics.

A 'smart weapon' is another example where the application of RFID implants could increase the security of a mobile artefact. On 14 April 2004 VeriChip announced a partnership with gun maker FN Manufacturing to produce a police gun with an RFID reader embedded. The objective of the partnership was to produce a gun that could not be fired by anyone but the authorised individual for that particular gun. A digital signal would unlock the trigger when the scanning device inside the handgun identified the authorised police officer. Failure to authorise would render the gun useless (for more information on this application, please refer to the case study presented at the conclusion of this chapter).

There are examples of individuals who have been implanted with RFID tags, which were originally manufactured for industry or supply chain purposes, and are equipped with cryptosecurity features.[5] These implants have been used to gain access to cars or to log into a computer.

3.5 Smart Environments

In the Ambient Intelligence (AmI) vision, people will be surrounded by intelligent interfaces that are embedded in a range of objects. These smart environments will be capable of recognising and responding to the presence of different individuals in a seamless, unobtrusive and often invisible way. RFID implants could provide the interface between people and the smart environments. Let us take the example of the car: the car's RFID reader would read a person's ID, recognise that she has permission to drive the car, open the door, adapt the seat height and position the mirrors. Similar applications could also be thought of for other smart environments, like homes. People have already volunteered to have RFID tags implanted to experience ambient intelligence environments, which adds some credibility to this as a concept.

---

[5] Graafstra 2007, p 18.

3.6 Other Potential Long-Term Applications

Examples of Location Based Services (LBS) include emergency services (location of emergency calls by the fire brigade, paramedics and police officers, for example) or local information services (finding the nearest restaurant, ATM, petrol station). Combining RFID implants with location-based systems may deliver new personalised services for people on the move. The communication range of RFID implants is limited to about 0.5 m (10–15 cm with a hand reader), but an intermediate device, such as a mobile phone equipped with an RFID reader, could provide a connection between the RFID implant and the network. The user would activate the interface device to request a service via the wireless network and the RFID implant would provide the user's identification.

RFID implants as replacements for externally attached devices used to localise or register the movements of prisoners has also been discussed by a number of commentators. An RFID tag, implanted or not, cannot be utilised as a global tracking device without a substantial infrastructure being put in place at prohibitive cost. However, future implant technology may incorporate transmitter modules (using e.g. ultra-wideband) for applications such as tracking children or prisoners.

The security issues relating to RFID in general have been well documented, especially those concerning the cloning of the device.[6] However, the permanent and physical link between an RFID tag and a person makes RFID implants more susceptible to privacy risks than any other kind of contactless tokens.[7] These will be further investigated in the Chap. 5.

Appendix

Case studies: The safe gun and access control for special rooms

Bernhard Anrig and Claude Fuhrer

Case study one: The safe gun
The 'Smart Gun' or 'Personalized Gun' is a concept gun that aims to reduce the misuses of guns by children and/or felons by using RFID tags or other proximity devices, fingerprint recognition or magnetic rings. For the moment, only magnetic devices are readily available.[8] The goal of the smart guns is to prevent the misuses of the weapon in events such as, for example, teenager suicides, unlawful homicides, and also prevent a stolen gun being used against a police officer during an operation.

---

[6] Halamka et al. 2006, p 699.

[7] Rotter 2008, p 70.

[8] http://en.wikipedia.org/wiki/Smart_Gun. Accessed 19 August 2011.

There are a range of different technologies that have been used to prevent the use of a gun by an 'unauthorised' person. One such technique is to implant an RFID tag under the skin of the owner of the gun. The grip of the gun embeds an RFID tag reader which can identify if the actual user of the gun is the authorised person for that particular gun. If not, then the trigger is blocked from releasing and it is not possible for the gun to be fired. VeriChip[9] is one of the prominent providers of this system.

Before using RFID implants, industry had tried to identify the authorised user of the gun by using a magnetic ring or a bracelet. Unfortunately, rings and bracelets can quite easily be duplicated. This drawback appears to be more important than the advantages of having a 'removable' identification system.

VeriChip (now called PositiveID[10]) has developed a tag approved by the US FDA. This tag has to be scanned with a given frequency and then it answers by transmitting a unique 16 digits number which can then be used to identify the tag and hence the owner of the tag. Considering that the world population is less than 10^{10} human beings, this 16 digits number allows to have about 10,000 different IDs for every person on earth. The use of this VeriChip technology has raised quite some controversies about its implications on users' privacy as the tag cannot be deactivated. This means that any time and place, a person bearing such tag can be identified, without consent, and even without being informed of the identification operation.

It is interesting to note that both gun-rights groups (like the National Rifles Association (NRA[11]) for example) and gun-controls groups (like the Violence Policy Center[12]) have criticised this technology.[13] While the former fears that the technology would restrict their right to own and use a firearm, the latter fear that the use of smart guns could provide arguments to 'legalise' a wider dissemination of guns in jurisdictions such as the United States.

One may also question the reliability of such a technology. Like every control technology, one will expect some 'false positive' identification; that is an unexpected authorisation of the use of the gun, as well 'false negative' identification, i.e. the gun refuses to be activated by its legal owner.

Deactivating the RFID tag

Normally, when an RFID tag has been implanted, it remains active until its removal or its complete destruction. There are different (and quite creative) ways to deactivate RFID tags. For example, one can mechanically destroy the tag with a knife or a cutter. It is also possible to destroy it electronically, for example by putting the object into a microwave oven for a couple of seconds.

---

[9] http://www.wired.com/science/discoveries/news/2004/04/63066. Accessed 19 August 2011.

[10] http://www.positiveidcorp.com. Accessed 19 August 2011.

[11] http://home.nra.org. Accessed 19 August 2011.

[12] http://www.vpc.org/aboutvpc.htm. Accessed 19 August 2011.

[13] http://www.wired.com/politics/law/news/2002/04/52178   and   http://www.vpc.org/fact_sht/smartgun.htm. Accessed 19 August 2011.

These methods are obviously not applicable in the case of RFID tags used for identification of weapon's owners. On the one hand, it is difficult to apply microwave heating to an implanted tag; on the other hand, the technology used must allow to temporarily disable an RFID tag, since once destroyed, you cannot reactivate the tag when the officer returns to work.

There are two problems to solve here. The first is to make the falsification of the tag more complicated than the simple transmission of a serial number; the second is to allow a police officer not to be identified as such when not in service. The use of a challenge-response authentication protocol may be employed to solve the first of these problems. An authentication mechanism could prevent duplication of tag and enhance the security of use. The second problem, namely deactivating a RFID, is more complicated. One can imagine that if the reader transmits a certain code to the tag, the tag could enter into a 'sleeping mode', only to be reactivated if the right sequence is subsequently transmitted to the tag. This implies that the tag has the necessary computing capabilities to do so and not only transmits some static serial number. Another approach is to give the user more possibilities of controlling the access to the RFID tag using different sensors.[14] However, while these ideas seem to work well for RFID tags on cards, corresponding applications to implanted RFID tags may need future work.

Dynamic grip recognition

Another technology, developed by the New Jersey Institute of Technology, is based on 'Dynamic Grip Recognition'.[15] To achieve this, the grip of the gun includes a set of sensors (such as, for example, mechanical pressure sensors). The measurements made by these sensors are continuously analysed and the trigger is released only if the profile established by the collected data corresponds to the reference data of the authorised user of the gun. But, to do all this processing, the firearm needs to have embedded in it a whole computer with some form of power supply. The user would need to be able to check at any moment the charge of the battery; the gun should also automatically inform the user that the battery level has reached a critically low level. One must ask, however, what would happen if the weapon is not able to protect its authorised user? A high-capacity battery is definitely an advantage, but being exempt from regularly recharging increases the risk of the user forgetting to check the battery's charge.

Another aspect of this technology that should be considered is the ratio of false negative recognition. The prototype showed a false rejection ratio of 1:100; that is the trigger was not released when the gun was in the hand of its authorised user one in every one hundred, or 1%, of the time. For daily operation this failure rate is

---

[14] Marquardt et al. 2010a, b.

[15] http://www.weaponsblog.org/entry/smart-gun-with-dynamic-grip-recognition-system/ and http://www.popsci.com/scitech/article/2005-12/smart-shooter and http://www.servamus.co.za/index.php?option=com_content&task=view&id=363. Accessed 19 August 2011.

too high and would need to be drastically reduced to, say, 1:10,000. For weapons used in the context of law enforcement, it is conceivable to add a mechanism to bypass the biometric recognition. This would allow, for example, a police officer whose weapon is unavailable or faulty to use the weapon of one of his colleagues.

Case study two: Access control for special rooms
In 2004, the Mexican Attorney General's office implanted 18 of its staff members with the VeriChip[16] implant in order to control access to a secure data room.[17] The room contained the country's database on criminals (the anti-crime information center). Given the importance of such data it is not too difficult to understand why the government and those within the country's law enforcement agencies were interested in controlling and monitoring access to such a room and the resource contained within. But what about the autonomy and informed consent of those individuals who were implanted with the tag? Did they really have the choice to accept or refuse the device?

One could argue that passive tags only provide a unique identification number and this number may be copied or duplicated. To prevent this kind of abuse, engineers have developed different techniques based on the 'normal behavior' of the tag owner. These techniques, already known in intrusion detection systems for computers, check for statistical anomalies in the activity of a user.'[18] If the behaviour of a user deviates from the expected behaviour then an alert will be generated. This requires fine-tuning of the technology in order to find a compromise between sufficient security and annoyance.

It is possible that having an implanted RFID device could make a person more vulnerable to attack. If a RFID reader identifies some people as 'chipped', one can imagine that these personalities, may be 'of interest' for criminal organisations. If someone has an RFID tag implanted, it may mean that this person has some specific responsibilities and then that one can obtain some reward (or ransom) if this person is robbed or kidnapped.

About the bibliography of this use case: Searching documentation about this use case may provide some surprises. There are not a lot of articles published on this subject (or at least there is not a lot of documentation indexed by major search engine's). Most of the available papers have been written by opponents of this system. To convince the reader that the implantation of RFID tags into humans is bad, many papers use religious argumentations. For example, one may often read that an RFID tag is 'the mark of the beast'.[19] Although religious views are important, one should be careful when using sources based on theses arguments.

---

[16] http://www.positiveidcorp.com/

[17] http://noidchip.free-forums.org/mexican-attorney-general-vt8.html. Accessed 21 April 2011.

[18] Thamilarasu and Sridhar 2008.

[19] http://www.ridingthebeast.com/articles/verichip-implant. Accessed 21 April 2011 or http://www.orwelltoday.com/readerverimob.shtml. Accessed 21 April 2011.

References

Graafstra A (2007) Hands on. IEEE Spectr 44(3):18–23

Halamka J, Juels A, Stubblefield A, Westhues J (2006) The security implications of VeriChip cloning. J Am Med Inform Assoc 13:699–707

Marquardt N, Taylor A, Villar N, Greenberg S (2010a) Rethinking RFID: awareness and control for interaction with RFID systems. In: Proceedings of the ACM conference on human factors in computing systems—ACM CHI'2010. ACM Press. http://grouplab.cpsc.ucalgary.ca/grouplab/uploads/Publications/Publications/2010-RethinkingRFID.CHI.pdf. Accessed 19 Aug 2011

Marquardt N, Taylor A, Villar N, Greenberg S (2010b) Visible and controllable RFID tags. In: Video Showcase, DVD proceedings of the ACM conference on human factors in computing systems—ACM CHI'10. (Atlanta, Georgia), ACM Press. http://grouplab.cpsc.ucalgary.ca/grouplab/uploads/Publications/Publications/2010-VisibleControllableRFIDTags.CHIVideo.pdf. Accessed 19 Aug 2011

Perakslis C, Wolk R (2006) Social acceptance of RFID as a biometric security method. IEEE Technol Soc Mag 25:34–42

Rotter P (2008) A methodological framework for the assessment of security and privacy risk for RFID systems. IEEE Pervasive Comput 7:70–77

Thamilarasu G, Sridhar R (2008) Intrusion detection in RFID systems. In: Proceedings of the Military Communications Conference, MILCOM08, pp 1–7. http://202.194.20.8/proc/milcom08/milcom08/pdfs/1882.pdf. Access 19 August 2011

Chapter 4
Restoring Function: Application Exemplars of Medical ICT Implants

Ryszard Tadeusiewicz, Pawel Rotter and Mark N. Gasson

Abstract Medical devices such as cardiac defibrillators and pacemakers used to restore heart rhythm and cochlear implants to restore hearing have become well established and are widely used throughout the world as a way in which to improve an individual's well-being and public health more generally. The application of implantable technology for medical use is typically 'restorative', i.e. it aims to restore some deficient ability. Notably, these sophisticated devices form intimate links between technology and the human body. Recent developments in engineering technologies have meant that the ability to integrate silicon with biology is reaching new levels and implantable medical devices that interact directly with the brain are becoming commonplace. Keeping in step with developments of other fundamental technologies, these types of devices are becoming increasingly complex and capable, with their peripheral functionality also continuing to grow. Data logging and wireless, real-time communications with external computing devices are now well within their capabilities and are becoming standard features, albeit without due attention to inherent security and privacy implications. This chapter explores the state-of-the-art of invasively implantable medical technologies and shows how cutting edge research is feeding into devices being developed in a medical context. Here, the focus of the analysis is on four technologies-pacemakers and cardiac defibrillators, cochlear implants, deep brain stimulators and brain computer interfaces for sight restoration.

R. Tadeusiewicz · P. Rotter (✉)
AGH University of Science and Technology, Krakow, Poland
e-mail: rotter@agh.edu.pl

R. Tadeusiewicz
e-mail: rtad@agh.edu.pl

M. N. Gasson
School of Systems Engineering, University of Reading, Berkshire, UK
e-mail: m.n.gasson@reading.ac.uk

M. N. Gasson et al. (eds.), *Human ICT Implants: Technical, Legal and Ethical Considerations,*
Information Technology and Law Series 23, DOI: 10.1007/978-90-6704-870-5_4,
© T.M.C. ASSER PRESS, The Hague, The Netherlands, and the author(s) 2012

Contents

4.1 Pacemakers and Cardiac Defibrillators

While both pacemakers and cardiac defibrillators, pager-sized devices implanted in the chest, are used to treat abnormal heart conditions, they have notably different applications. The primary purpose of a pacemaker is to maintain an adequate heart rate, either because the heart's native pacemaker is not fast enough, or because the electrical system controlling the heart has failed. The pacemaker continually monitors heart activity, and when an abnormal heartbeat pattern occurs, it can stimulate the ventricle of the heart via an implanted electrode lead with a short electrical pulse. An implanted cardiac defibrillator (ICD), on the other hand, is used in patients who are at risk of sudden cardiac death due to ventricular fibrillation. By detecting cardiac arrhythmia, it can revert both atrial and ventricular arrhythmias as well as perform biventricular pacing by delivering electrical stimulation. Some devices combine a pacemaker and cardiac defibrillator in a single implantable unit.

Pacemakers and cardiac defibrillators typically consist of a sealed, battery powered, sensor-laden pulse generator, several steroid-tipped wire electrodes that connect the generator to the heart muscle, and a custom ultralow-power microprocessor, typically with about 128 Kbytes of memory for telemetry storage. The device's primary function is to sense cardiac events, execute therapies, and store measurements such as the history of the device's operation. Modern devices communicate wirelessly and are programmable to allow the cardiologist to select optimum parameters for individual patients post implantation. Device manufacturers also produce at home monitors that collect data from the implanted devices through the wireless channel and relay it to a central repository, accessible to doctors via an SSL-protected website.

4.2 Cochlear Implant Technology

Prosthetic devices for hearing loss vary significantly depending on the condition being treated. Typical hearing loss can be repaired by simple hearing aids; these are often simple sound amplification devices. These were first developed in the early part of the twentieth century, and for a number of decades the technology has

Cochlear Implant

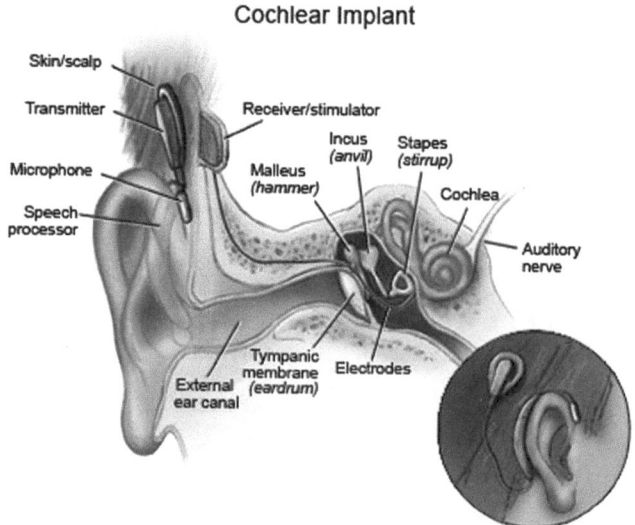

Fig. 4.1 General scheme of the cochlear implant–a kind of artificial ear. *Source* http://
kidshealth.org/parent/medical/ears/cochlear.html#cat138

been well received. However, if deafness is very profound or occurs from birth,
restoration cannot be done using the typical external apparatus for the deaf and
instead invasive devices known as cochlear implants are utilised. The method of
hearing restoration based on cochlear implants is a standardised medical procedure
and according to the United States (US) Food and Drug Administration (FDA),
approximately 188,000 people worldwide had received implants as of April 2009.
The cochlear implant is probably the most popular and the most widely used type
of implant that interfaces directly with the human nervous system.

As shown in Fig. 4.1, biological and artificial (technological) elements work
together to make the system function. An implanted part of the system is located
inside the patient's body, under the skin covering the skull behind the ear. This
makes contact with the internal biological structures of the ear through a small
canal in the skull bone. Signals from the implant are transmitted to the cochlea, a
biological sound receptor which is part of the internal ear and gives its name to the
devices. The majority of the device is located outside the skull with communication through live tissue achieved by wireless technologies.

The general structure of a cochlear implant is designed as a technological
analogy of the functioning healthy human ear. A microphone is used to collect
sound wave signals, replacing the function of the eardrum. These signals are
processed and passed wirelessly inside the head where a stimulator is used to
excite cells within the cochlear to create the perception of sound. In normal
hearing, complex mechanical properties of the basilar membrane, a part of the
inner ear located inside the cochlea, cause sounds of different frequencies to excite
hearing receptor cells located in the Corti organ (the so-called hair cells) located in

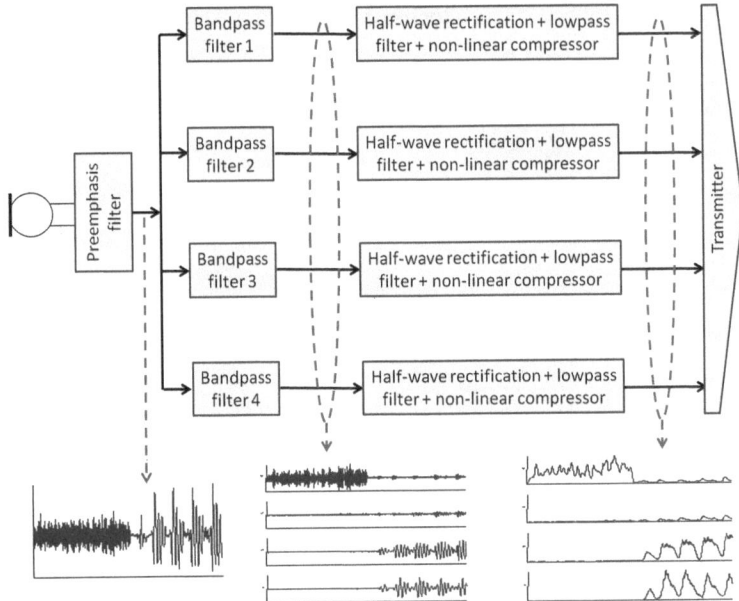

Fig. 4.2 Typical signal waveforms at selected points during sound processing

different places inside the cochlea. There is a precise mapping between sound frequencies and locations of the excitation inside the Corti organ: high frequency sounds excite receptor cells located near the base of the cochlea and low frequency sounds excite receptor cells in the upper (apical) part of the cochlea named the helicotrema. Excitation of receptor cells in the Corti organ create an electrical signal which is then registered by bipolar neural cells (spironeurons) belonging to the spiral ganglion. The nerve fibres form the auditory nerve, which pass information to the brain.

Taking into account the steps described above, the designers of the cochlear implant were first required to divide registered sound signals into segments assigned to particular frequency bands. This signal processing is performed in the speech processor unit. The speech processor first prepares signals from the microphone by means of 'pre-emphasis filtering'. A pre-emphasis filter is used for attenuation of strong and not very informative components in speech below 1.2 kHz. Pre-emphasis filtering is followed by multiple processing channels, where sound energy is measured in particular frequency bands. Because internal ear electrical stimulation cannot be performed very quickly, the output of every bandpass filter in the speech processor is subjected to full-wave rectification and low-pass filtering. The envelope signals extracted from the processed bandpass filters are additionally compressed with a nonlinear mapping function. This transformation must be performed prior to the stimulation because of a big

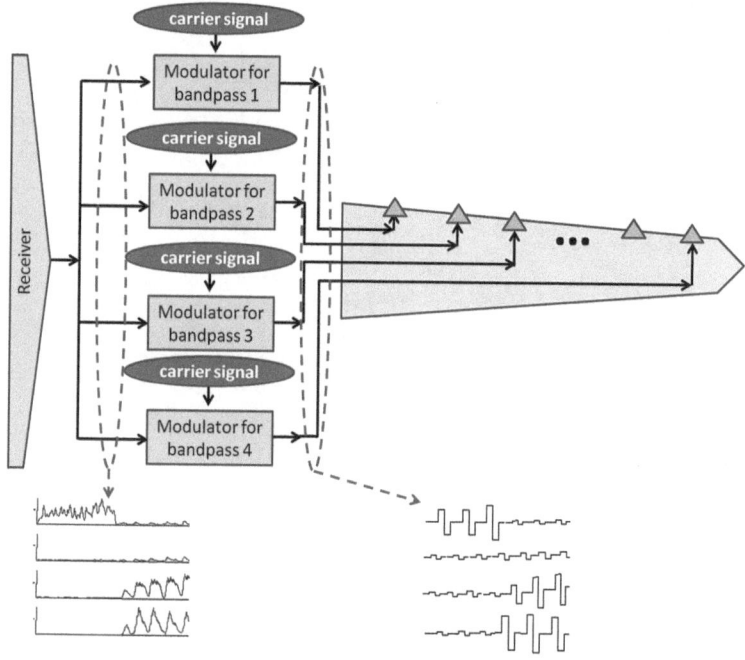

Fig. 4.3 Modulated waveforms are generated to trigger electrical stimulation of specific electrodes in the cochlear implant

difference between the wide dynamic range of sound in the environment (up to about 100 dB) and the narrow dynamic range of electrically evoked hearing (about 10 dB). Such prepared signal (so-called auditory signal envelopes) are used for the control of the stimulation process. Typical waveforms of signals in selected points of the speech processor structure are shown in Fig. 4.2.

Compressed envelope signals are transmitted by the wireless communication system to the internal part of the cochlear implant system, located under the skin. Signals received by the implanted receiver are used to control the intracochlear stimulation. The output of each channel (related to the corresponding bandpass filter in the speech processor) is used for modulation with a special carrier signal, as shown by Fig. 4.3. Each channel is directed to a single electrode in an array structure coiled inside the cochlea internal canal (the scala tympani) which is only a few millimetres in length. Low-to-high channels are assigned to apical-to-basal electrodes, to mimic at least the order-if not the precise locations-of biological frequency mapping in the cochlea.

Cochlear implants are an effective technology for hearing restoration for the profoundly deaf. For patients with badly damaged, but existing hearing, it is possible to observe disadvantages connected with two-way sound perception, i.e. natural hearing using, for example, bone sound transmission to the cochlea and the

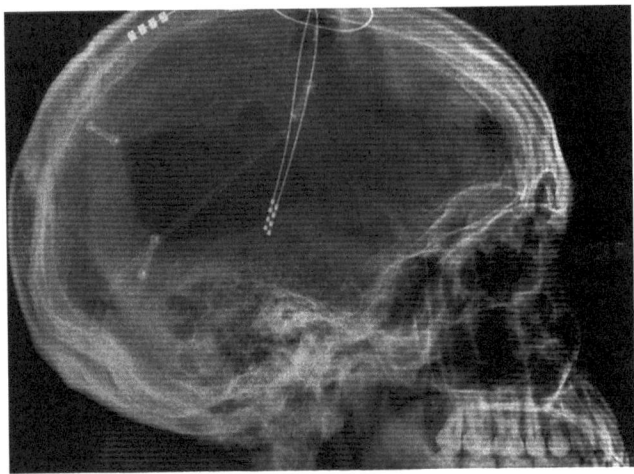

Fig. 4.4 A lateral X-ray of the head of a 38-year-old showing two Deep Brain Stimulation leads implanted in the subcortical thalamus area

additional path formed by a cochlear implant. For these patients it is necessary to further develop cochlear implant systems which can take into account the effect of residual natural hearing and improve hearing in frequency bands where physiological elements of the ear are not functionally sufficient. Indeed, next generation cochlear implant development must be directed towards advanced signal processing. For the enhancement of speech perception, especially in noisy environments, more advanced methods of sound signal processing and stimulus delivery must be developed which more closely resemble normal cochlea function.

4.3 Deep Brain Stimulators

In the neurotechnology field, deep brain stimulation (DBS) is a surgical treatment involving the implantation of a medical device that sends electrical impulses to specific parts of the brain. DBS in selected brain regions has provided remarkable therapeutic benefits for otherwise treatment-resistant movement and affective disorders such as chronic pain, Parkinson's disease (PD), tremor and dystonia.

Recently there has been a resurgence of interest in the surgical treatment of movement disorders such as PD. This is because of the disabling side effects of long-term treatment with L-dopa, a chemical precursor to dopamine which can cross the blood–brain barrier and metabolise in the brain to address insufficient dopamine levels, thought to be a primary cause of PD. Also many movement disorders, such as multiple system atrophy or dystonia, do not respond to dopaminergic treatment at all. A limited range of DBS systems have been made commercially available and are now in clinical use despite their significant cumulative costs.

Deep brain electrodes are routinely implanted into the thalamus, pallidum or subthalamic nucleus to alleviate the symptoms of Parkinson's disease, tremors of multiple sclerosis and dystonia. In pain patients, electrodes are implanted into the sensory thalamus or periventricular/periaqueductal grey area (see, for example, Fig. 4.4). The depth electrodes are externalised for a week to ascertain effect prior to internalisation. A control unit and battery is implanted in the chest cavity and the electrode connections internalised after this time if good symptom relief is realised, at a cost of around £12,000.

Typical DBS systems consists of: the electrode implanted in the brain with its tip positioned within the targeted area, an insulated wire extension which is passed under the skin of the head, neck and shoulder and connects the electrode with the neurostimulator, also containing the battery, which is implanted under the skin near the collarbone. Once the stimulator is fully connected it delivers a continuous electrical pulse to the targeted brain area via the electrode. At present DBS is used to stimulate deep brain structures continuously at high frequencies (typically 100–180 Hz for movement disorders and 5–50 Hz for pain). Such high frequency DBS is probably effective because it takes command of the local networks and prevents them from relapsing into the slow synchronous cycles that may cause the symptoms of the disorder. The corollary of this is that when entrained to continuous deep brain stimulation the basal ganglia neurons are probably unable to perform their normal functions. The general therapeutic stimulation parameters of the pulse have been derived primarily through trial and error. This has been possible in disorders such as Parkinson's disease as the effects of DBS on symptom alleviation manifest quickly—in contrast to other disorders.[1] The ability to optimise stimulation parameters based on the changing pathophysiology of the disorder targeted is vital for the application of DBS technology. As such, the devices utilise wireless communication channels in an almost identical fashion to those of pacemakers and cardiac defibrillators.

The success of DBS as a treatment for the symptoms of movement disorders, combined with an improved understanding of the pathophysiologic basis of neuropsychiatric disorders has now seen renewed interest in the application of DBS for these conditions.[2] Despite the increasing successful clinical use of DBS, its mechanisms of operation are still unclear. Further to this, continuous electrical stimulation does not allow the targeted brain area to engage in normal function. In addition, patient quality of life is adversely affected by repeating operations (every ~3 years) to replace stimulator batteries as a direct result of continuous stimulation. This is costly in terms of time and money and currently severely limits the number of people able to undergo this highly successful treatment.

---

[1] McIntyre et al. 2004, p 40.

[2] Wichmann and DeLong 2006, p 197.

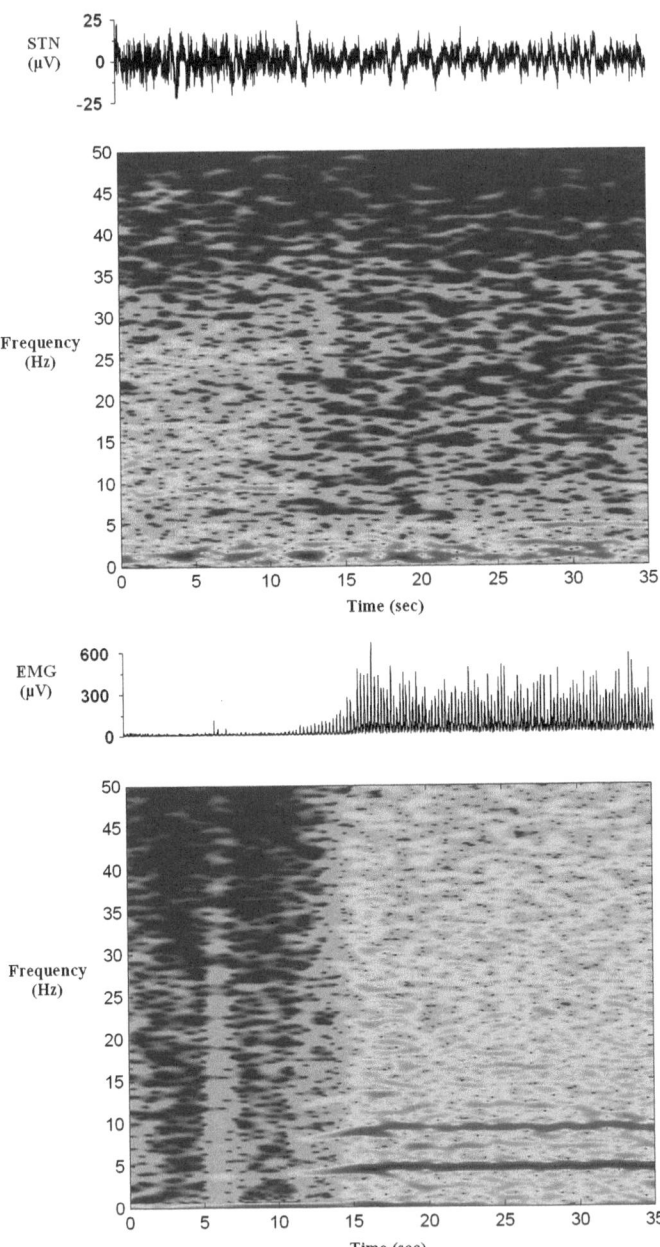

Fig. 4.5 STN LFPs (1st plot) and forearm extensor surface EMGs (3rd plot) during the onset of resting tremor, together with their respective STFT spectrograms. These show the potential for invasively recording and decoding brain activity, and using this information to artificially manipulate it

Because of this, neurophysiologists, bioengineers and signal analysts, are currently collaborating to develop a device that is able to utilise predictive brain activity to trigger the stimulator. Such devices, in a similar but more complex way to cardiac defibrillators, monitor the brain activity through the electrodes, and respond to abnormal patterns of activity with electrical stimulation to abort it. The effectiveness of noncontinuous, on-demand stimulation has been shown in epilepsy,[3] where brief pulse stimulation can terminate epileptiform after discharges. It is likely that the additional complexities involved by detecting such events will see computationally more complex devices being utilised verses current cardiac devices.

The depth electrodes are externalised for a week to ascertain effect prior to internalisation. A unique opportunity exists to record the local field potentials (LFPs) from the nucleus via the deep brain electrode and to correlate these to electromyograms (EMGs) recorded simultaneously from affected muscle groups to better understand what changes in brain activity cause the undesirable symptoms. Since the electrode is available for monitoring LFPs during normal stimulator operation if continuous stimulation is not utilised, then it becomes feasible to utilise analytic paradigms which show that it is possible to detect signal changes in the target nuclei prior to the event of tremor.[4] Figure 4.5 shows the raw recordings and spectrograms of recordings made simultaneously from the subthalamic nucleus (STN) within the brain and from the surface of the forearm extensor muscle in a patient with Parkinson disease, over the onset of resting tremor. The STN LFP spectrogram shows a significant suppression in the range of 10–30 Hz (the beta band), appearing around the 12–13th second followed by increased power at the tremor frequency of 4.6 Hz appearing at the 18th second. Meanwhile, rhythmic tremor bursts appeared in the surface EMGs at the 18th second in association with two strong narrow bands at the tremor and double-tremor frequencies and a general increase in activity across the entire displayed frequency range in the EMG spectrogram. The power of the STN LFPs in the beta band (10–30 Hz) decreased significantly approximately 3–5 s preceding the power increases at the tremor frequency in both STN LFPs and surface EMGs over the onset of resting tremor.

This shows the feasibility of invasively recording brain activity, and using these recordings as predictors of further function that can be modified by electrical stimulation. As with many developments in this area, this work is based within a medical context. However, it demonstrates the transferable knowledge available from disparate domains which is relevant to the human enhancement field of research.

---

[3] Motamedi 2002, p 836.

[4] Gasson et al. 2005.

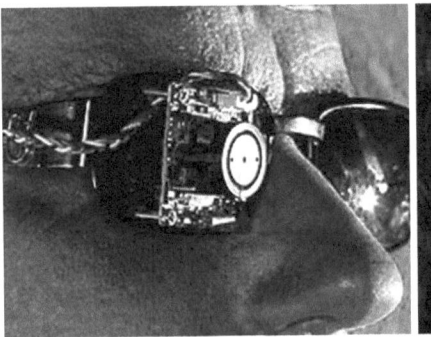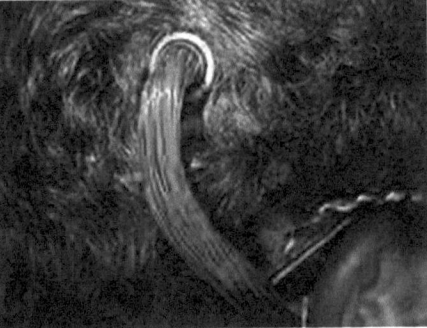

Fig. 4.6 Dobelle artificial vision system: the camera mounted in glasses and cables which send the image to the brain. *Source* http://www.wired.com/wired/archive/10.09/vision_pr.html

4.4 Brain Computer Interfaces for Sight Restoration

The quality of cameras became sufficient to deliver a high quality image several years ago. While in theory such digital cameras could replace biological eyes and even enhance human sight, establishing a connection between a camera and the brain is much more difficult than constructing the sensor itself.

Experiments with camera-brain interfaces began in the early 1970s.[5] A grid of electrodes was implanted into a blind patient's visual cortex and was used to produce phosphenes—flashes of light perceived by the patient at different visual positions. Until the beginning of the twenty-first century, continual progress was made primarily due to the development of tools for image processing and not the implant interface. By upgrading signal processing hardware and software Dobelle achieved good results using electrodes implanted over 20 years earlier: the patients were able to detect obstacles and read 15-cm letters from about 2 m.[6] Several years later the first commercial Brain Computer Interface (BCI) implant for sight restoration was used for the first time.

The system shown in Fig. 4.6 is similar to the first BCI implants: the image from the camera is transmitted to a portable computer, which is carried on the patient's belt. Image processing removes the background and this is then further processed before being passed to electrodes implanted in the visual cortex. In 2000, Dobelle stated that 'it is unlikely that patients will be able to drive an automobile in the foreseeable future'.[7] Two years later a patient with a second-generation Dobelle implant was able to drive slowly around a parking area shortly

---

[5] Dobelle and Mladejovsky 1974, p 553.

[6] Dobelle 2000, p 3.

[7] Dobelle 2000, p 3.

after an operation to fit the system. The image perceived is, however, still very simple and does not enable normal driving.

Recently progress in this area has slowed. The major problems are that the image is perceived as a set of bright points on a black background (a starry night sky effect) and the field of view is very narrow (tunnel vision). In other words, the perceived image is of extremely low resolution. The image seen by the patient does not facilitate colour vision or depth perception. The cost is also much higher than any other technology designed to help the blind and there are important health concerns.

There are a number of medical risks related to invasive BCI, most significant of all is infection. This is the reason why the best known blind volunteer had connections from electrodes to the camera removed. The results achieved in the past decade are very interesting from a scientific point of view but it is questionable whether in light of high risk and very high cost this justifies implants. Coulombe, Sawan and Gervais proposed invasive BCI with reduced infection risk, where the electrodes are connected to the computer wirelessly.[8] But other technologies may offer comparable (if not better) mobility for the patients at much lower cost and without medical risks, like example, vOICe technology which changes the image into the sound.

References

Coulombe J, Sawan M, Gervais J-F (2007) A highly flexible system for Microstimulation of the visual cortex: design and implementation. IEEE Trans Biomed Circuits Syst 1(4):258–269

Dobelle WH (2000) Artificial vision for the blind by connecting a television camera to the visual cortex. ASAIO J 46:3–9

Dobelle WH, Mladejovsky MG (1974) Phosphenes produced by electrical stimulation of human occipital cortex, and their application to the development of a prosthesis for the blind. J Physiol (Lond) 243:553–576

Gasson MN, Wang SY, Aziz TZ, Stein JF, Warwick K (2005) Towards a demand driven deep-brain stimulator for the treatment of movement disorders. MASP2005, 3rd IEEE international seminar on medical applications of signal processing, London, 3–4 Nov 2005, pp 83–86

McIntyre CC, Savasta M, Walter BL, Vitek JL (2004) How does deep brain stimulation work? Present understanding and future questions. J Clin Neurophysiol 21(1):40–50

Motamedi GK (2002) Optimizing parameters for terminating cortical afterdischarges with pulse stimulation. Epilepsia 43(8):836–846

Wichmann T, DeLong MR (2006) Deep brain stimulation for neurologic and neuropsychiatric disorders. Neuron 52(1):197–204

---

[8] Coulombe et al. 2007, p 258.

Part II
Technical Challenges
of Human ICT Implants

Chapter 5
Passive Human ICT Implants: Risks and Possible Solutions

Pawel Rotter, Barbara Daskala and Ramon Compañó

Abstract Despite the advantages that the use of Radio Frequency IDentification (RFID) technology entails, privacy concerns stemming from its mass deployment causes concern both in society and the academic community. Although non-implantable passive RFID devices in general remain more advanced than implantable, glass capsule types, both generally have fundamental technological limitations that mean they are susceptible to similar issues. Indeed, the security and privacy implications of RFID for a variety of applications have been well explored; however, the use of them inside the body serves to further aggravate some of the known issues. Whilst the numbers of people with such implanted devices are still small, the commercialisation of this technology means that they could become commonplace and so these concerns are paramount. In this chapter, these issues are described, and it is argued that controls and measures can be developed and engineered now in order to minimise such risks. In addition, the medical risks specific to implantation of these devices is explored in light of recent research.

P. Rotter (✉)
AGH University of Science and Technology, Krakow, Poland
e-mail: rotter@agh.edu.pl

B. Daskala
European Network and Information Security Agency (ENISA), Heraklion, Greece
e-mail: Barbara.Daskala@enisa.europa.eu

R. Compañó
European Commission Joint Research Center, Institute for Prospective Technological Studies (IPTS), Seville, Spain
e-mail: Ramon.Compano@ec.europa.eu

M. N. Gasson et al. (eds.), *Human ICT Implants: Technical, Legal and Ethical Considerations,*
Information Technology and Law Series 23, DOI: 10.1007/978-90-6704-870-5_5,
© T.M.C. Asser Press, The Hague, The Netherlands, and the author(s) 2012

Contents

5.1 Privacy Risks of Radio Frequency IDentification Applications

Despite the advantages that the use of Radio Frequency IDentification (RFID) technology entails, privacy stemming from its mass deployment causes concern both in society and the academic community. In July 2011, a Google search reported 55 million pages related to RFID, from which over 25% was related to privacy. One of the main reasons why RFID tags pose special privacy concerns is the fact that they usually respond to any reader that interrogates them. This behaviour, combined with the fact that humans cannot sense the transmission used to perform the reading, makes RFID tags an ideal target for imperceptible eavesdropping. Moreover, tags do not maintain a history of readings. Thus, any unauthorised reading cannot be detected at a later time, reinforcing the undetectability of these transmissions. This promiscuity, together with the limited computational, storage and communication abilities of the tags, raises privacy concerns unique to the use of RFID technologies—especially in relation to human usage.

Secondly, most RFID tags' responses include a *unique identifier*. Thus, even if any extra data a tag transmits is cryptographically protected, and no information can be extracted from it, the device (or person) carrying the tag can still be tracked. Further, it is not uncommon that a person carries more than one tag. This creates a constellation of identifiers that can be linked amongst themselves and to the carrier. This situation increases the risks associated with traceability by making it easier to do and more accurate.

RFID tags and RFID constellations potentially pose a threat to what is commonly referred to as 'location privacy'. Here, the main goal is to prevent other parties from learning the location and further movements of a person. By placing covert RFID readers in strategic places it is possible to monitor movements. For instance—and as discussed in Chap. 3—a company could place these readers inside its buildings, allowing the tracking of its employees. This information could be used by the management to make inferences about individual employees' productivity rates. It would not be necessary to make these readers secret as RFID cards are often used to enforce access control to restricted areas. The information provided by these readers could already be used to extract information about employees' behaviour.

Moreover, when personal information and RFID tags can be linked, a person may not only be tracked by an unknown party but may also be subjected to further privacy invasions. For instance, when a payment with credit card is carried out next to an RFID reader, the identity of the payer, and the constellation of tags that individual carries may be directly linked. It may be the case that these tags reveal information about the products they are attached to, such as the type of product: clothes, mobile phone, medicine, etc., or even more detailed information: aspirin, syrup, etc. This data may then be used to infer personal information about the carrier (relating to, for example, illnesses or shopping preferences). The surveillance of a tag (or a group of tags) associated to an individual may leak personal information as well as reveal the behaviour or intentions of that person. In a retail environment the fact that a customer stops for a long time in front of a shelf may reveal their purchase intentions. Readers can be placed to detect this behaviour and the information can be exploited by shop assistants to increase the sales or to perform customer profiling.

It is important to realise that this association of identities and tags made in shops, companies and so on is inexact as they only take into account the 'association'—by this we mean that a person is related to a tag but not the opposite operation where a person breaks this association (losing, throwing away, selling, etc. the tagged item). This 'eternal' association in databases may become a threat if the item is used afterwards for any malicious activity, for which the original owner may be incriminated.

Up to this point we have only considered commonly denominated *personal privacy threats*. However, RFID tags attached to goods also involve corporate *data security threats*, where the risk is shifted from the leakage of personal information to business sensitive information (espionage, marketing, etc.). As explained earlier, RFID tags are hailed as the future of product and merchandise identification. The fact that RFID technology does not necessarily require close physical proximity between reader and scanned objects, nor a line-of-sight between them, makes it ideal for inventory management. However, these characteristics might be a double-edged sword, by allowing third parties to spy on a business at different steps in the production. The threats stemming from RFID use in the production of goods are not limited to information theft and misuse. Finally, using RFID as a core technology in the supply chain makes it susceptible to denial of service attacks, as radio frequency signals can be easily jammed to stop or at least complicate the full production process.

5.2 Specific Privacy Risks of RFID Implants

The previous section canvassed some of the major privacy concerns related to the use of RFID technology. RFID implants, which are based on RFID technology, are bound to share the same concerns; however, given the nature of RFID implants, we would argue that they pose a number of additional risks.

In general, one could say that the privacy risks affecting people with RFID implants differ slightly from the risks stemming from the use of other contactless tokens. To start with, none of the so-called *corporate data security threats* apply anymore because implants are not part of massive production processes. However, the aforementioned threats to privacy still concern implant users. Moreover, the fact that RFID implants are *permanently* linked to a person make them especially susceptible to privacy invasions and, in some cases, this persistent connection may even worsen the situation, exposing the RFID users to physical danger.

When talking about RFID implants, the undiscriminating behaviour of RFID tags with respect to any reader becomes crucial. Such implants answer any reader's request—usually with a message containing its unique identifier (we recall that encryption is not necessarily a solution for this problem). This identifier can, in this case, be linked with absolute certainty to a physical person; this is contrary to other tokens where this association cannot be ensured. For instance, people may share access cards or car keys, which then helps to 'hide' the actual owner of the token. However, when the token is implanted, the anonymity set of possible owners is reduced to one. This facilitates traceability.

Permanent identifiers associated to people make constellations of RFID tags heighten the potential threats relating to privacy. As in the previous case, the fact that the RFID tokens may be exchanged prevents inferences about identifiers appearing in two or more constellations. On the contrary, if the identifier of an implanted tag is part of two constellations, it may be used as evidence that in both cases the same physical person was detected.

These drawbacks are also very important when speaking about location privacy, or further inferences about private information that can be derived from a tag constellation (including, for example, use of a product/medicine). Again, the fact that a tag is uniquely related to a person makes it easier to extract information and use it afterwards. Nevertheless, the permanent association of tag/user may become an advantage in scenarios where users need to prove their presence in some place, or to avoid false incriminations stemming from the misuse of contactless tokens. In such scenarios, however, one has to consider that these tags can be cloned, and as such this may not be of any benefit.

Finally, the use of implants for authentication purposes, for instance, in an access control environment, may result in physical damage to the user. An attacker may want to extract the implant from the body, or even tear it apart if necessary. It has been argued that the solution to this problem lays in either allowing easy cloning of the devices[1] or limiting its use to identification.

5.3 Some Medical Concerns of RFID Implants

The US Food and Drug Administration (FDA) have paid special attention to the potential medical risks posed by implanted RFID tags as part of their regulatory function associated with approving new medical devices. In reviewing the

---

[1] Halamka et al. 2006, p 699.

application for the VeriChip RFID implant, one can assume that the FDA identified no significant risk/s in relation to the device by virtue of the fact that it gave clearance for the commercialisation of the implant. The FDA did, however, point to several *potential* medical issues associated with the implant: "adverse tissue reaction; migration of the implanted transponder (…), failure of implanted transponder; failure of inserter; failure of electronic scanner; electromagnetic interference; electrical hazards; magnetic resonance imaging incompatibility; and needle stick".[2]

On 8 September 2007, an article titled "Chip Implants Linked to Animal Tumors" reported on several studies—dating to the mid 1990s—which suggested a link between cancer and laboratory rodents who had been injected with an RFID implants. The quoted percentage of rodents which developed tumours differed between the studies, with some authors suggesting 1%, while another cited a figure closer to 10%.[3] The head of the Department of Health and Human Services, which oversees the FDA, became a member of the VeriChip board after the approval which raised questions about the objectivity of the approval process. The FDA would not reveal which studies were reviewed before they took a decision to approve the implant, or whether the abovementioned reports were taken into consideration. VeriChip claimed that the company was not aware of them.

It has been argued that observations presented in the cited studies do not imply any risks for people implanted by RFID.[4] The main argument put forward by the commentator was that tumours among laboratory rodents can often be caused by just the injection itself. During the past 15 years, for example, millions of dogs and cats have been injected with RFID implants and only one case of cancer at the site of the implant was reported. It is however probable that many were not examined in this way, or reported.

So far, no health problems related to implants were reported among 2,000 people injected by VeriChip. It is also worth noting that the reports claiming high medical risks were authored by a privacy activist who has opposed RFID technology very strongly, including on religious grounds. Indeed, the results found in the literature are not consistent and rather conflicting. Even based on an apparently alarming study,[5] the probability of causing cancer appears to be very small, so the study recommends leaving the implant in injected pets. For people already implanted, it suggests leaving the decision whether to remove it or not to the individual. Given the risk/benefit trade-off of having an RFID implant, clearly more research in this area is needed.

---

[2] "Class II Special Controls Guidance Document: Implantable Radiofrequency Transponder System for Patient Identification and Health Information." U.S. Department of Health and Human Services, Food and Drug Administration, Center for Devices and Radiological Health, December 10, 2004.

[3] Studies on larger population indicated a lower percentage (1% among over 4,200 mice, while 10% among 177 mice).

[4] Wustenberg 2007.

[5] Albrecht 2007.

5.4 Discussing Some Possible Solutions

Privacy advocates and researchers have extensively studied the privacy risks associated with RFID technology and have, to date, made a number of suggestions aimed at mitigating such risks. It is important to note here is that to address the concerns and mitigate the risks posed by RFID implants, we would need to consider the following:

5.4.1 Implementation at Various Levels

Controls and measures would need to be considered and implemented at many different levels including, for example, technical, organisational, social and policy levels. Enforcing only technical measures—although useful in their own right—would not adequately address the totality of issues associated with the technology. Rather, citizens must also be kept aware and educated about the risks entailed, so that they are then able to make informed choices. This is clearly understood if one considers, for example, the case of an organisation with a strong password policy. Although this is undoubtedly best practice, employees often choose very strong passwords, which they then write down on a post-it note sticking it on their computer screens. This undermines the technological measure adopted, and does not actually offer the desired information security level the technical measure would normally provide.

There are a range of possible technical controls that have been already identified to mitigate RFID security and privacy risks. RFID tags may, for example, be put into a 'sleep' mode. When applying this solution, one has to take care to introduce a mechanism to ensure that only authorised readers can 'wake up' the tag. This could be achieved by means of a PIN or through more complex methods.[6] 'Sleeping' strategies could be applied to ICT implants, but implant bearers would need to take care of the PIN. However, it is important to note that if PINs are not secretly stored, the method is not effective anymore, and to some degree this additional level of user interaction defeats some of the purpose of the device. Some scholars have proposed the blocker tag.[7] In this scheme, tags contain what is called a modifiable 'privacy bit'. If this bit is set to '0', the tag behaves normally, thus responding to any reader that interrogates it. If the bit is set to '1', the tag is considered as private. Once tags are categorised, a 'blocker tag' will prevent private tags from being scanned. While this solution could be applied to RFID implants, its security is not completely assessed and blockers cannot guarantee full protection.[8] Finally, privacy may also be protected through measures that do not prevent undesired scanning of the tags but make the detection of these scans

---

[6] Stajano and Anderson 1999.
[7] Juels et al. 2003.
[8] Rieback et al. 2005.

easier.[9] However, these techniques rely on the goodwill of providers and customers and cannot be seen as an ultimate solution for privacy invasions derived from the use of RFID tags.

In terms of 'softer' controls, it is important that citizens are kept aware of the risks and most importantly are also educated on the use of these technologies and its impact. Citizens may be willing to accept a certain level of risk in relation to their privacy if, for example, they consider that the risks—real or perceived—are outweighed by the perceived benefits. However, and wherever possible, they should be given the opportunity and ability to make their own decisions.

5.4.2 Being Proactive and Managing Risk Appropriately

Controls and measures should also be developed in a proactive manner: this is normally referred to as 'privacy-by-design'. The idea behind this is simple: identification of security and privacy risks of a technology or application a posteriori, leading to identification and implementation of corrective measures as an afterthought, is neither a cost-effective procedure nor an efficient one. On the contrary, identification of information security and privacy controls at the design phase of the application/technology development is bound to address these concerns more appropriately and cost-efficiently.

A recent and rather successful example towards this direction is the Privacy Impact Assessment framework for RFID applications that has been proposed by the industry and endorsed by the EU Article 29 Working Party in February 2011.[10] This was performed in the context of the European Commission's Recommendation on RFID in which the Commission called upon the Member States to provide inputs to the Article 29 Data Protection Working Party for development of a privacy impact assessment framework to be used in the deployment of radio frequency identification (RFID) applications.[11] In our view, this approach is a move in the right direction. An important feature and benefit of this is the introduction of a systematic risk management practice when dealing with this type of risk.

Today identification implants based on passive RFID are still in their infancy. There is, however, recognition in the public and private sectors that the technology has significant potential application across numerous sectors. Passive RFID devices may complement other identification methods and provide easy way of accessing services and devices. A limitation of the technology is that, at present,

---

[9] Juels and Brainard 2004; Garfinkel 2002.

[10] "Privacy Impact Assessment Framework for RFID applications" available at: http://ec.europa.eu/information_society/policy/rfid/index_en.htm.

[11] European Commission 2009.
   http://ec.europa.eu/information_society/policy/rfid/documents/recommendationonrfid2009.pdf

there is not sufficient support for security of data and protection of privacy. This, together with concerns about health and ethical issues, is a significant obstacle for further deployment. The future success of RFID implants, especially in relation to their use in humans, will depend on workable and innovative solutions which address these concerns.

References

Albrecht K (ed) (2007) Microchip-induced tumors in laboratory rodents and dogs: a review of the literature 1990–2006. http://www.antichips.com/cancer. Accessed 15 Apr 2011

Garfinkel SL (2002) An RFID bill of rights. Technol Rev 150(8): October

Halamka J, Juels A, Stubblefield A, Westhues J (2006) The security implications of VeriChip cloning. J Am Med Inform Assoc 13:699–707

Juels A, Brainard JG (2004) Soft blocking: flexible blocker tags on the cheap. In: Proceedings of the 2004 ACM workshop on privacy in the electronic society, pp 1–7

Juels A, Rivest RL, Szydlo M (2003) The blocker tag: selective blocking of RFID tags for consumer privacy. In: Proceedings of the 10th ACM conference on Computer and communications security, pp 103–111

Rieback M, Crispo B, Tanenbaum AS (2005) Keep on blockin in the free world: personal access control for low-cost RFID tags. Security Protocols Workshop, pp 51–59

Stajano F, Anderson RJ (1999) the resurrecting duckling: security issues for Ad-hoc wireless networks, Security Protocols Workshop, pp 172–194

Wustenberg W (2007) Effective carcinogenicity assessment of permanent implantable medical devices: lessons from 60 years of research comparing rodents with other species

Chapter 6
Implantable Medical Devices: Privacy and Security Concerns

Pawel Rotter and Mark N. Gasson

Abstract The technical issues associated with human ICT implants are many and varied. While several of these are associated with technology operating in a hostile environment, there are many others which centre around our lack of understanding of the human body, and in particular the brain with its inherent complexities. This has meant that we are limited in our ability to interface the silicon of technology with the biology of the body in truly meaningful ways. However, as research continues to develop solutions to these barriers, the systems which result are potentially vulnerable to technical issues such as security and privacy which are familiar from other mainstream application paradigms. Building systems which address these issues from the outset rather than as an afterthought is an important design strategy. However, with core functionality at the forefront of the designers minds, already there is evidence that medical devices exist which fail to address these concerns. Here, we outline some of the core technological issues which are already beginning to pervade medical human ICT implant devices.

P. Rotter (✉)
AGH University of Science and Technology, Krakow, Poland
e-mail: rotter@agh.edu.pl

M. N. Gasson
School of Systems Engineering, University of Reading, Berkshire, UK
e-mail: m.n.gasson@reading.ac.uk

M. N. Gasson et al. (eds.), *Human ICT Implants: Technical, Legal and Ethical Considerations*,
Information Technology and Law Series 23, DOI: 10.1007/978-90-6704-870-5_6,
© T.M.C. Asser press, The Hague, The Netherlands, and the author(s) 2012

Contents

6.1 Introduction

There are several substantial differences between Radio Frequency IDentification (RFID) implants (as discussed in Chap. 5) and ICT implantable medical devices (IMDs):

- IMDs are not optional—usually the patient has no choice. On the other hand, identification implants (RFID tags) are implanted mostly for convenience or even for fun. Although it has been argued that RFID implants are lifesaving devices, clearly alternative solutions for identification exist such as wearable tokens (these need not necessarily be electronic),
- insecure implementation of IMDs may provoke much more serious threats than in the case of identification implants. A malicious attack on IMDs if successful may directly threaten the life of the patient by for example changing the implant's parameters or triggering the device, e.g. causing defibrillation,
- the device itself, as well as communication with it, is much more complex than in the case of RFID implants.

Below we present privacy and security risks related to the use of implantable medical devices. All risks listed below are related to the wireless communication between the IMD and external world.

Potential risks to the *privacy* of an individual include:

- Unauthorised scanning of people in order to detect the presence and type of medical implant
- unauthorised reading of the device ID number, which can then be used for tracking of people
- unauthorised reading of the patient's personal data, which are often saved in the device's memory, and/or
- unauthorised reading of medical data collected by the device, which gives insight into patient's health state.

There are several risks related to *security* including:

- Malicious modification of firmware or data stored in a device's memory
- malicious modification of the device's configuration parameters

- stopping the device from operating. This can be achieved by sending the instruction code which stops the device, or by a denial-of-service attack, and
- triggering a device into action which may threaten a patient's health and life, for instance by triggering defibrillation.

Researcher's have already demonstrated that some existing clinical devices are at risk of serious security flaws,[1] and the security issues surrounding the next generation of clinical devices have also been highlighted.[2]

The risks to privacy and security are context-dependent. In every particular case, only some items from the list above are relevant and specific analysis of risks should be performed for every individual application.

6.2 Security/Accessibility Trade-Off

There is a trade-off between security and accessibility of the device, which needs to be seriously considered. For the sake of functionality, which can be crucial for a patient's life, especially in emergency situations, the communication with the device should be easy and free of limitations. For example, if an unconscious patient is brought to an emergency department in a foreign country, it is likely that the personnel will not be able to follow advanced authorisation procedures to obtain control over the device.[3] On the other hand, requirements for privacy and security cannot be disregarded. The possibility of unauthorised access to the device, reading of the data and manipulation of the device by an unauthorised person is a serious risk to a patient's safety and privacy. Therefore, there is a need to find a compromise between security measures and accessibility requirements.

6.3 Means of Protection for IMDs

Basic means of protection against unauthorised access to IMDs are:

- Access control: authorisation of specific people (patient, patient's doctor) or entities (ambulance staff) to perform specified operations on the device. Different operations may require different levels of authorisation, for example manual control over the device (including entering testing mode) is the most potentially dangerous.

---

[1] Halperin et al. 2008, p 129.

[2] Denning et al. 2009, p 1.

[3] Halperin et al. 2008, p 129.

- requirement of authorisation via a secondary channel, for example by using near field communication to initialise the data transmission.
- a set of measures against denial-of-service attack, in particular against the overflow of the device memory, the draining of the battery power and the blocking of communications.

Also some other measures can be applied for IMDs[4]:

- Notification to the patient when the device exchanges data with an external reader through a secondary channel (e.g. vibration), and
- using an intermediate device for communication, which could be embedded e.g. in a smart phone, watch or belt. Then short-range communication between the IMD and the intermediate device could use light-weight encryption and authentication, while strong security could be applied for the exchange of data between intermediate device and external devices.

While such solutions would not be technically difficult to implement, it is currently the case that commercial products lack even basic access control, and so are significantly vulnerable to attack. Security through obscurity may work in the very short term, but other mechanisms need to be implemented.

6.4 Conclusion

It is clear that a range of human ICT implants are being developed and utilised, and they are bringing with them their own security and privacy implications. This is in the main because these devices are being developed without due consideration to these important aspects, in many case without even basic access control being implemented. However, in light of the personal data which can be contained within, and the possibilities for identification and utilisation by unauthorised people, it is clear that consideration needs to be made to these areas. This is clearly better addressed while the technology is still relatively immature.

References

Denning T, Matsuoka Y, Kohno T (2009) Neurosecurity: security and privacy for neural devices. Neurosurg Focus 27(1):E7
Halperin D, Heydt-Benjamin TS, Ransford B, Clark SS, Defend B, Morgan W, Fu K, Kohno T, Maisel WH (2008) Pacemakers and implantable cardiac defibrillators: software radio attacks and zero-power defences. In: IEEE symposium on security and privacy, pp 129–142

---

[4] Halperin et al. 2008, p 129.

Part III
A Social, Ethical and Legal Analysis of Human ICT Implants

Chapter 7
Carrying Implants and Carrying Risks; Human ICT Implants and Liability

Arnold Roosendaal

Abstract Human information and communication technology (ICT) implants can have a major impact on the people carrying them. Unfortunately, there may also be negative consequences, resulting in damages. To claim for damages to be recovered, it has to be clear who can be held liable. Is this the healthcare practitioner who placed the implant, the manufacturer, the person who programmed or updated the software, or does the patient carry all the risks himself? This chapter discusses medical liability and product liability in the context of human ICT implants. The debate on the distinction between therapy and enhancement is briefly touched upon and the importance of causation is discussed. Challenging issues specific for human ICT implants, such as security of the ICT components, is introduced. It appears that the complexity of the implants as well as the myriad of people involved in the development, programming, and placing of the implants lead to possible difficulties in determining who can be held liable for damages and the extent of information duties of manufacturers and healthcare practitioners.

Arnold Roosendaal is a consultant at Fennell Roosendaal Legal Research. The author gratefully acknowledges Professor Carla Sieburgh (Radboud University Nijmegen) for her valuable comments on earlier drafts of this chapter.

A. Roosendaal (✉)
Fennell Roosendaal Research, Tilburg, The Netherlands
e-mail: arnold@fennellroosendaal.nl

M. N. Gasson et al. (eds.), *Human ICT Implants: Technical, Legal and Ethical Considerations,*
Information Technology and Law Series 23, DOI: 10.1007/978-90-6704-870-5_7,
© T.M.C. Asser Press, The Hague, The Netherlands, and the author(s) 2012

Contents

7.1 Introduction

A central issue in the application of modern technologies concerns the question of liability. What to do if something goes wrong and, more particular, who can be held responsible for damages arising from this 'wrong'? In the case of human information and communication technology (ICT) implants there can be two forms of liability at stake: medical liability and product liability. This chapter discusses the question of liability with regard to ICT implants. At the European Union (EU) level, there is only limited specific regulation of human ICT implants,[1] dating from the 1990s and without particular attention for liability issues. This implies that a connection has to be found in 'normal' medical liability and product liability legislation.

Medical law has largely focused on the duty to take reasonable care and skill of the health care provider. There has to be reasonable care of the health care provider so that if there is an absence of effect this does not necessarily constitute a shortcoming in the fulfillment of a contractual agreement or an unlawful act. This line of reasoning extends so that it cannot be said that there is always material damage or harm, a requirement for liability, for the patient if the intended effect is not reached. From a liability perspective this raises the question whether someone can be held liable for the absence of an effect or the manifestation of another effect than intended. The patient may consider this to be damage, but whether this damage has to be paid for depends on whether someone can be held liable for this damage. In particular, in cases where there is no clear distinction between therapy or enhancement purposes of the implant this can be problematic. Another difficulty lies in causation as a necessary requirement for liability. To be able to claim damages it has to be proven that the damages are the result of the act of the health care provider concerning the implant. In other words, the act of implanting needs to be the cause of the damages that occurred. This relationship can be very difficult to prove depending on the facts in question due to, among other things, the

---

[1] Council Directive 90/385/EEC of 20 June 1990 on the approximation of the laws of the Member States relating to active implantable medical devices.

complexity of the human body. This may lead to several possible causes of a certain effect.

In Sect.7.2 of this chapter, medical liability will be discussed, including different approaches that exist in Europe and a brief discussion on the distinction between therapy and enhancement. Subsequently, in Sect. 7.3 product liability will be discussed. Having the two relevant forms of liability, the concept of causation and the way this influences liability in cases concerning human ICT implants will be described in Sect. 7.4. Finally, in Sect. 7.5 some points of discussion and concluding remarks will be presented.

7.2 Medical Liability

Medical liability is mainly regulated at national levels. Some European legislation is available, but this is merely related to good clinical practices[2] and medical devices[3] in general, without directly addressing liability issues. Obviously, non-compliance with these Directives may be an indication for a lack of reasonable care and skill, but the legal implications depend on the national regulations. The national approaches concerning medical liability can be distinguished in three forms, which will be discussed below.

Nevertheless, the European Group on Ethics (EGE) has indicated that "implantable devices for medical purposes should be regulated in the same way as drugs when the medical goal is the same, particularly as such implants are only partly covered by Council Directive 90/385/EEC."[4] This seems to be a feasible approach, in particular because more clarity is given by focusing on the aim of implanting a device. In this respect, it is important to notice that, in the EGE opinion, the aims are restricted to medical goals, therewith excluding enhancement purposes. Directive 90/385/EEC contains a vast number of guidelines and requirements concerning ICT implants, but this merely addresses the implant itself and not the health care provider who places the implant. Attention to this Directive will be paid later on when it concerns product liability.

Concerning medical liability three approaches can be distinguished in different European countries. Two main approaches are strict liability and negligence. The third form is the so-called no-fault approach.

Strict liability means that a health care provider can always be held liable for damages that occur after a medical act. The fact that a health care provider performs certain acts implies that he has to carry the risks. When liability based on

---

[2] Directive 2001/20/EC of the European Parliament and of the Council of 4 April 2001 on the approximation of the laws, regulations and administrative provisions of the Member States relating to the implementation of good clinical practice in the conduct of clinical trials on medicinal products for human use.

[3] Council Directive 93/42/EEC of 14 June 1993 concerning medical devices.

[4] European Group on Ethics 2005, p 35.

negligence is applied, it has to be proven that the health care provider did not perform his task correctly or completely. To prove negligence, a lot of difficult investigations may be necessary in order to show that the health care provider acted improperly. The complexity of medical acts can make this a complicated exercise. Several factors, such as medical conditions at the time of the medical act, awareness of possible complications, and the performance of the act itself have to be taken into account to provide a complete picture. These investigations cost a lot of money and can be an emotional burden for the patient. In order to solve this problem, some countries[5] have introduced the no-fault system. In this system, there are no investigations on negligence or fault and damages are covered from the insurances of the health care providers. In practice, this is less burdensome for victims, but almost similar to strict liability, since the health care provider has to pay damages (via his insurance). The only difference, though important, is that there is no statement that the health care provider is guilty of negligence or fault, or is held liable based on risk liability.

A complicating factor when it concerns ICT implants is that normal medical performance is connected to a duty of care. The health care provider has to take reasonable care and skill to reach a certain result. There is, however, no duty to achieve a result. Besides, it can be difficult to determine what the standard of care exactly is that has to be met in order to have fulfilled the duty of care. For instance, in The Netherlands the decision on the standard of care is made by a judge by checking whether the health care provider has exercised the level of care expected from a conscientious health care provider.

In the UK, the medical community decides on the standard, because the 'standards of reasonably competent medical men at the time' have to be applied.[6] So, standards may differ among countries and the way of defining the standard, as well as the elements taken into account during process, differs as well. In the case of ICT implants this might mean that the mere (careful) implantation is sufficient to avoid liability, regardless of whether the implant has the intended effect. This is different when the act of implanting the device as such was irresponsible due to medical indications in general or specifically related to the patient or due to characteristics of the implant itself. The health care provider always has to act as a professional.

Nevertheless, when human ICT implants are at stake this may be a difficult issue. First, there is the fact that these implants are relatively new and that for this reason the effects may not be clear. However, what kind of materials can be used in implants should be more common knowledge by now based on a relatively long history of implants.[7] In addition, before it is allowed to use an implant it usually has been subjected to a medical approval test. There are strict guidelines in the medical sector concerning approval processes of new applications. These can

---

[5] For instance France and in a slightly different form Belgium.

[6] Hondius 2010, p 12.

[7] Think, for instance, of screws and pieces of plate metal that are used to support bones or joints.

concern drugs as well as methods and instruments and materials. Independent commissions have to give their approval before a new medical product may be launched onto the market. The effects pointed at here are related to the acceptance of the implant by the human body and the possible connection to tissue or other parts of the body.

A second difficulty relates to the effects of implants in a physical or neurological way. Here, the functionality of the implants and in particular the ICT component is at stake. In some cases, such as with a pacemaker, the aim will be to influence the functionality of the body. May a pacemaker be something commonly accepted and used, it is not difficult to imagine sophisticated and more questionable applications of ICT implants. RFID chips can be used for payments,[8] but other implants may trigger the brain to steer the functioning of the body or to alter certain functions. The difficulty lies in the distinction between therapy and enhancement and whether this distinction influences questions concerning liability. Traditionally, there are three medical functions: diagnosis, advice, and treatment. It is questionable whether enhancement falls under this traditional approach, because there is no ordinary medical diagnosis that has to determine what exactly a patient is suffering from. It can, however, also be said that the advise part is more important in cases of enhancement, in particular, when it concerns very recently developed implants.

From the opinion[9] of the EGE it can be derived that it makes a clear distinction between therapy and enhancement by pointing at the medical purpose of implantable devices. Therewith the EGE seems to be willing to exclude enhancement purposes from the regulatory framework as it exists for drugs. Still, the questions concerning liability remain, since they are not covered in the other regulations. The EGE also points out that currently 'non-medical implants in the human body are not explicitly covered by existing legislation.'[10] This lack of specific legislation means that general legislation is applicable. Thus, the above mentioned systems as applied at national levels are leading and a duty of effort is the primary focus point.

In the light of the distinction between therapy and enhancement there may, however, still remain difficulties. In particular, when it concerns an implant with enhancing functionalities the scope of the requested effort may be unclear. Is research concerning the expected result part of the effort? And does a health care provider have to weigh alternatives? Due to the newness and the enhancing function of these implants, the expectations of the medical act may become a major factor. But does this possibly influence liability when these expectations are not met? Probably, more support can be found in product liability.

---

[8] Like in the Baya Beach Club where visitors can have a chip implanted in their upper arm. This chip can be used as an identifier in order to get access to the club. Inside the club, payments can be made with the chip, because it is connected to an individual account of the visitor.See, for instance, Gossett (2004) Paying for drinks with wave of the hand. http://www.wnd.com/?pageId=24179. Accessed 1 July 2011.

[9] European Group on Ethics 2005, p 5.

[10] European Group on Ethics 2005, p 35.

7.3 Product Liability

Product liability is the liability for damage caused by a defective product. This is the basic approach as laid down in the European Product Liability Directive.[11] In some jurisdictions, damages resulting from the defective product, such as compassionate allowances, can also be claimed. Concerning (active) medical implants, Directive 90/385/EEC[12] sets out strict essential safety requirements. The implant devices,

> must be designed and manufactured in such a way that, when implanted under the conditions and for the purposes laid down, their use does not compromise the clinical condition or the safety of patients.[13]

The general safety requirements also contain a weighing of risks against benefits: 'Any side effects or undesirable conditions must constitute *acceptable risks* when weighed against the performances intended.'[14] Annex 1 furthermore lists a number of requirements regarding design and construction of the implantable devices. When the use of nonmedical implants becomes more frequent even stricter rules may be required since safety risks are not counterbalanced by effects on health.[15]

In cases where health is at stake, the risks taken will be weighed against the potential benefits on the health of the patient. This weighing will determine the boundaries of medical care. If there is no medical necessity, the weighing cannot be based on this, which implies that other factors may be decisive. These factors, or the process of weighing the risks, can be regulated by law. At least it is clear that the implant has to be a safe product. Damages resulting from a defective implant can be serious, since a human carrying the implant is involved. The manufacturer of the product can be held liable for damages. An important step to limit liability is to meet legal requirements and standards. If there are technical standards, the manufacturer of a device should make sure that he meets these requirements and keeps track of the state of the art technology. This can be of importance in showing that the manufacturer of a device did everything he could to make a safe product.

A specific question concerns the effect of an implant. When there is no effect or another effect than intended, this basically might mean that the product is a defective product. Depending on the type of implant this is, however, not that straight-forward. There are two complicating factors. One is that no effect or another, or smaller, effect than intended can find its cause in the implant as such, but

---

[11] Council Directive 1985/374/EEC of 25 July 1985 on the approximation of laws, regulations and administrative provisions of the Member States concerning liability for defective products, OJ L 210, 7 August 1985, 29.

[12] Council Directive 90/385/EEC of 20 June 1990 on the approximation of the laws of the Member States relating to active implantable medical devices, OJ L 189, 20 July 1990, 17.

[13] Annex 1 of the Directive, under 1.

[14] Annex 1 of the Directive, under 5 (emphasis added).

[15] De Hert and Mantovani 2010, p 113.

can also have to do with other factors related to the body of the person who received the implant. Another complicating factor concerns the question of what to do when there is a defective product. Product liability law provides for replacement of defective products. But in the case of an implant this may be very difficult or even risky. It implies another medical treatment and the implant can be attached to body tissue. Thus, for a manufacturer it may be very difficult, if not impossible, to replace a defective implant. In some cases, this might even lead to ongoing damages.

A difficult issue that comes up concerns the question whether the lack of any effect can also be seen as damage, in particular, when the implant was intended for enhancement purposes. The same question may hold when the occurring effect is different than the intended effect, depending on the exact effect in practice. Are damages necessarily negative consequences or can maintenance of the status quo also be considered as damage? The latter is normally the case. However, it is very well possible that the effects of an ICT implant cannot always be guaranteed. When this is the case, this should be made clear beforehand. When, though unfortunate, the effect does not take place or only in a limited manner this can probably not be considered to be damage and the manufacturer of the implant cannot be held liable.

As to the liability implications for manufacturers of products, the life cycle of adoption of a technology may be a good indication. When a device is first brought onto the market there may be predictable implementation snags that may heighten the liability risk of the manufacturer (or health care provider). However, usually the manufacturer can rely on a state of the art defense, that negative effects were not known at the moment and that the manufacturer was not supposed to know these negative consequences either, to exclude liability. After the initial period, the application and workings of the devices become clear and more common, but also creates opportunities for errors. In this period, the liability risk will be quite stable. Finally, long-term effects of implants can include potential shifts in the legal standard of care that has to be applied.[16]

An interesting aspect in the case of certain ICT implants, such as RFID tags, which contain a unique identification number or have any form of network capability, is that liability is probably not limited to the implantable device itself. There can be a connection to a database or network that also has to be safe. The database and the software are in some way extending the scope of the implant. This might mean that mistakes in the database or software can also raise liability issues. Obviously, the same counts for mistakes in software running in the implant itself. The specific characteristics of ICT implants may thus introduce difficulties concerning liability, since several people may be involved in the development, manufacturing, programing, and implantation process.

---

[16] Compare: Mangalmurti et al. 2010.

7.4 Causation

Liability requires causation. In order to be able to hold someone liable for damages it has to be proven that the damages are the result of the act of the health care provider. Well-known examples of this are the cases where health care providers leave instruments behind in the belly of a patient. Another example may be that the health care provider, when placing the implant, brings damage to a healthy organ of the patient. To claim damages based on product liability, the damages have to be the result of the defective product. To prove that damages are the result of the act of the health care provider may be very difficult in cases where the damages take the form of side effects in the functioning of the body or in the body itself. If the body itself gets damaged, for instance if there is an infection, it still may be unclear whether this is due to the implant or whether something else went wrong during the implantation surgery, or whether it has actually nothing to do with the implant at all. In any case, how causation is dealt with differs between various European jurisdictions. In some jurisdictions, causation is little more than a *condition sine qua non* test, of a mainly physical nature. In other jurisdictions, causation is the main criterion for deciding on whether or not someone will be held liable.[17]

A distinction can be made between specific causation and general causation. General causation occurs in cases where the case can regularly repeat itself. Specific causation occurs in a traditional accident case, which is specific and probably itself. Here, evidence concerning causation is generated in a single incident, after which the evidence degrades.[18] Witnesses forget details or are no longer available, and physical evidence becomes less reliable or gets lost. In contrast, 'evidence about general causation typically improves over time.'[19] There are two reasons for this: 'first, the *data* available for studies on general causation improves over time,' and second, 'even if the amount of available data [...] remains constant, *the studies* that collect and analyze the data improve over time.'[20]

An example of causation can be seen in the *Implanon implant* case. Even though this case does not concern ICT implants it is a case in point, which can shed some light on the difficulties surrounding causation. Implanon is an implant in the form of a small tube that acts as a 3-year contraceptive. The implant was introduced worldwide and in clinical trials it demonstrated a lower failure rate than existing forms of contraception.[21] In Australia, the manufacturer of Implanon, Organon, developed a training program for medical practitioners to introduce the contraceptive. However, in Australia, as well as in numerous other countries where

---

[17] Hondius 2010, p 15.
[18] Cheng 2003.
[19] Cheng 2003, p 326.
[20] Cheng 2003, p 327 (emphasis in original).
[21] Wenck and Johnston 2004.

Implanon had been introduced, it appeared that there were relatively a lot of unintended pregnancies. It was not that there was one clear reason why the implant did not work properly, but possibilities included

> failure to insert the implant, incorrect timing of the insertion, and implant failure, possibly, in some instances, because of interactions with other medications.[22]

In Australia this resulted in new insights and approaches toward risk assessment of newly introduced medical products with a high-risk factor.

In the Netherlands, cases concerning the unintended pregnancies have been brought before the court.[23] In one case, a woman got pregnant after the implantation of Implanon. Several tests showed that the implant was not present in the body at the moment the woman had been pregnant for 22 weeks already. It was unclear whether the implant had been there and failed to work[24] or whether the implant was just not implanted or not correctly, resulting in expulsion. It was also questionable whether the health care practitioner had checked the presence of the implant after insertion. In any case, the result was a mal-performance or non-performance of the doctor, and the doctor, or—based on specific laws—the hospital as responsible for all its personnel, should be held liable.

It seems that liability risks for defective products are carried by the manufacturer of the implant, while all risks concerning the surgery, information provision, proper functioning, and correct implantation are carried by the health care practitioner. This implies that the health care practitioner has a quite extensive task in learning about the functioning of the implant, risk factors, possible limitations on the functioning of the implant, and the technical process of placing the implant. Most likely, this will be the same in cases of ICT implants.

Some specific concerns remain, however, in relation to ICT implants. ICT implants have an additional component as compared to other implants, which is the software component. Software carries a specific risk. Programming the implant is a precise task, which has to be carried out without any mistakes. A small mistake, even a typo in a line of code, can give a different meaning to the program or obstruct the overall functioning of the program. If the program steers an ICT implant that influences the functioning of the human body or a small part of it, any mistake can have major consequences.

Another risk related to ICT is that the software can probably be hacked and changed from a (small) distance, or that a chip may be erased by an electromagnetic field. Similar risks occur when the implant has to be checked or updated every once in a while by connecting, physically or wireless, to a computer or network. For instance, this may be the case with implants that are programmed to

---

[22] Wenck and Johnston 2004, p 117.

[23] Rb. Alkmaar, 11-02-2004, 60177/HA ZA 02-519, LJN: AO3453; Rb. Alkmaar, 24-05-2006, 60177/HA ZA 02-519, LJN: AX4831; Hof Amsterdam, 24-01-2008, 106.005.258, LJN: BC9815; Hoge Raad, 24-12-2010, 09/02579, LJN: BO4579.

[24] Pearl-index data show that, despite the use of Implanon, there can occur 0,03 pregnancies per 100 women years.

release a medicinal dose at fixed times to help patients suffering from Alzheimer's disease getting their medicines at the right moment. These doses for medication may change regularly, meaning that the ICT implant needs to be reprogrammed. ICT implants can have extensive functionalities, ranging from performing complex analyses, decision-making capabilities, storing of personal medical data, and automatic communication, remotely, and wirelessly:

> These features have provided improved care and quality of life for millions of patients, but they also have created a susceptibility to security breaches that could compromise the performance of such devices and the safety and privacy of patients.[25]

Patients should be aware of these risks, so health care practitioners should inform them properly and make sure that patients make an informed choice concerning the implant. However, manufacturers have a responsibility here as well, in the sense that they should make sure that devices meet security standards and that software is properly protected against intrusion.

7.5 Concluding Remarks

ICT implants raise important issues concerning liability. Only limited regulation specific for these implants is available, so most is regulated by general laws on liability, in particular on medical liability and product liability. Some things are still unclear. Does the distinction between implants for therapeutically purposes and for enhancement purposes make a difference when it comes to liability? Medical acts require the prior informed consent of the patient, but the exact effects of an implant may not be clear beforehand. In particular, when it concerns ICT implants that influence brain functioning or physical functions of the body. If a 'patient' chooses for an ICT implant for enhancement purposes and later loses control over his body and hurts another person, is this then still the result of a willful act of the person and can he, thus, be held liable? Or is this perhaps a case of product liability, because of the malfunctioning of the implant? Or is the health care practitioner liable for probably not informing the patient well about the risks (and uncertainties) of the implant? Or is the health care practitioner liable because of the fact that these enhancing implants are placed without strict medical necessity, which is also an ethical consideration? Or might there have been an error in the software of the implant, caused by an external factor? The relatively great variety of people involved in an ICT implant and the relatively huge impact if something goes wrong, make that liability questions may not be that easy to answer. In particular, causation may be difficult to prove. Besides, if damages imply removing or replacing the implant, this is very burdensome for the patient since another surgical operation might be necessary and just leaving the implant in the body may not even be an

---

[25] Maisel and Kohno 2010, p 1164.

option. A basic question is whether a health care practitioner is allowed to place an ICT implant without a medical need, while the implications of the implant are unclear but it is known that there may be certain risks related to the implant.

ICT implants can be connected to databases, either permanently or on a regular basis. In these cases the boundaries of the implant are unclear. Is the database part of the implant? And can temporary unavailability of the database be considered malfunctioning of the implant? Besides, databases, as well as the software running in an implant, can be vulnerable to attacks. Here, security, even though not elaborated upon in this contribution, is a specific issue that needs attention.

Experiences with ICT implants are limited and the implants can take many forms. They can be active or passive, therapeutic or enhancing, for body functioning or for brain functioning, etc. The diversity of applications and types makes it impossible to give a complete coverage in a short book contribution. However, the issues mentioned above and the background concerning legislation and lacunae in legislation, showed that many difficulties may arise and that there are still a lot of uncertainties surrounding liability in relation to human ICT implants. Whether legislation can completely solve all the issues is uncertain. However, some major dilemmas need attention, such as the distinction between therapy and enhancement, and can be covered in legislation. More clarity on where liability lies, with the health care practitioner or with the manufacturer of the implant, and to what extent, can be given as well. Providing clarity in this is a task for the legislator. A good start would be to identify the different forms of possible mistakes (medical, implant, software) and the related parties that can be held liable. Self-regulation by the sector may lead to a quest for excluding and shifting liability between the different actors involved, so this seems too difficult to provide clear solutions in such a complex field with such a variety of different actors involved. In any case, the complex interaction between medical care and ICT technologies, combined with a myriad of actors involved in the ICT implant development and production process makes it a challenging field for liability issues.

References

Cheng EK (2003) Changing scientific evidence. Minn Law Rev 88:315–352

De Hert P, Mantovani E (2010) The EU legal framework for the e-inclusion of older persons. In: Mordini E, de Hert P (eds) Ageing and invisibility. IOS Press, Amsterdam, pp 83–120

European Group on Ethics (2005) Opinion 20 on ethical aspects of ICT implants in the human body, 16 March 2005

Hondius E (2010) General introduction. In: Hondius E (ed) The development of medical liability. Cambridge University press, Cambridge, pp 1–26

Maisel WH, Kohno T (2010) Improving the security and privacy of implantable medical devices. N Engl J Med 2010:1164–1166

Mangalmurti SS et al (2010) Medical malpractice liability in the age of electronic health records. New Engl J Med 363(21):2060–2067

Wenck BCA, Johnston PJ (2004) Implanon and medical indemnity: a case study of risk management using the Australian standard. Med Law 181(2):117–119

Chapter 8
Implants and Human Rights, in Particular Bodily Integrity

Arnold Roosendaal

Abstract Human ICT implants can have various implications for human rights. In particular, the right to bodily integrity may be at stake due to the direct connection to the human body. In this chapter, the scope of the right to bodily integrity is discussed. A more detailed discussion on how the right is affected by implants is largely based on biotechnological implants. In this field, the distinction between therapy and enhancement, as well as implications of implants for the concept of the body and its integrity, has been largely debated. Subsequently, the right is approached with a focus on other, non-living implants. Specific emphasis is given on nanotechnological implants and information carriers. It also appears to be relevant whether the implant has an active or a passive functionality. Finally, some concluding remarks are made by pointing at additional challenges for human rights resulting from ICT implants.

Contents

A. Roosendaal (✉)
Tilburg Institute for Law, Technology, and Society (TILT), Tilburg University,
Tilburg, The Netherlands
e-mail: A.P.C.Roosendaal@tilburguniversity.edu

M. N. Gasson et al. (eds.), *Human ICT Implants: Technical, Legal and Ethical Considerations*,
Information Technology and Law Series 23, DOI: 10.1007/978-90-6704-870-5_8,
© T.M.C. ASSER PRESS, The Hague, The Netherlands, and the author(s) 2012

8.1 Introduction

Human implants are artificial parts, devices, or living organs which are implanted into the human body. The deployment of human implants has induced various implications for human rights. Human rights are meant to protect individuals and are often related to their dignity. The protection offered by these rights is essential for individuals to live a decent life. Therefore, human rights are often laid down in international treaties, such as the *Universal Declaration on Human Rights* (UDHR) or the *Charter of Fundamental Rights of the European Union.* The rights include, among others, the right to life and anti-discrimination.

One of the most crucial rights at issue in relation to implants is the right to bodily integrity,[1] and so, this contribution largely centres on this specific right, which may be described as having a twofold character:

1. It implies the right to prevent one's body from being harmed by others, and
2. also constitutes a right to do with one's body whatever one wants—a right to self-determination.

Implanting parts that do not originally belong to the body evidently implies a possible impact on bodily integrity. Often based on the UDHR—which dates from 1948—the right has only been included in Constitutions for a couple of decades.[2] Notwithstanding its relatively young legal status, the right may already be influenced by unforeseen developments in the field of technology. In this chapter, the right to bodily integrity will be discussed in relation to human information and communication technology (ICT) implants. However, since a lot of discussion concerning the right and implants took place in the context of biotechnological implants, these will be taken as an example in order to define the main points of discussion.

---

[1] For instance, as laid down in Article 3 of the Charter of Fundamental Rights of the European Union, dating from 2009.

[2] See for example, Article 2 para. 2 of the German Constitution and Article 11 of the Dutch Constitution; cf. also Chap. 2, Article 6 of the Swedish Instrument of Government (a part of the Swedish constitution).

8.2 Implants and the Human Body

Biotechnology, nanotechnologies and radio frequency identification (RFID) have become part of society and are gradually—but increasingly so—influencing our daily lives. These technologies can also be connected to or implanted in the human body. The purposes of implants can vary from strictly medical purposes, via aesthetic purposes, to human enhancement.

The development towards combining human beings with technological implants may have a great impact on social life and the concept of certain values, such as human dignity and privacy. The main research question at this point is how the constitutional right of the integrity of the human body is influenced by new technologies such as nanotechnologies, RFID or biotechnology. The hypothesis is that the concept of this right or the idea of the human body might be subject to slight changes when artificial parts become implanted into it. These artificial parts may still be considered as separate parts, which are implanted, but not part of the body itself. But what if a human organ is replaced by an artificial organ which is made by using biotechnology? In that case it is not a supplement *to* the body but a replacement in the body; in this case such an implant may consequently be considered as a part *of* the body.

If we accept this development, which has already begun, does this mean that this addition to the body is not violating the right of bodily integrity? And if it appears that bodily integrity is not considered to be violated by such implants, does it mean that the concept of this right is changing or has already changed? With regard to brain implants which have been employed to, for example, reduce the tremors in Parkinson's disease, the European Group on Ethics (EGE) expressed the opinion that,

> [t]hey show that ICT implants may influence the nervous system and particularly the brain and thus human identity as a species as well as individual subjectivity and autonomy.[3]

This remark suggests that the link with dignity and bodily integrity is very clear.

With respect to bodily integrity, another question would appear to mirror the question relating to which part of the body should be protected with integrity. As the functioning of the body can be expanded with external devices, the boundaries of the body may become increasingly unclear. The persons involved in experiments with implants and connections to external devices may experience the connected devices also as parts of their body. One of the most prominent researchers in the area of extending the body is Kevin Warwick,[4] who after implanting himself with an RFID implant, made the following comment:

---

[3] European Group on Ethics 2005.

[4] Professor at the University of Reading and the First self-proclaimed cyborg, see: http://www.kevinwarwick.com.

> The biggest surprise for me during the experiment was that I very quickly regarded the implant as being 'part of my body', a feeling shared with most people who have a cochlea implant or a heart pacemaker. In my case though, there was a computer linked to my implant and because the computer was making things happen, I very quickly became attached to it as well.[5]

Although the computer or other external device was not a physical part of Warwick's body, it was performing personal activities. Warwick, and indeed others, may therefore consider it as part of the body because the link is as direct as possible.

The question is whether this implies that an infringement on the device also infringes the right to bodily integrity, based on the restrictions brought to the linked person. So here is another criterion that may be used to determine whether the right to bodily integrity is under tension because of technological implants: what exactly does the person consider to be part of her body?

When using this criterion, problems may arise when individuals are connected to a network or series of networks. Nowadays, most people are connected to the Internet via their laptop or mobile phone. In a comparable way, when people carry implants, they can be connected to networks in order to expand their human capabilities; accordingly, people also become connected to each other. An implication is that different people may try to steer or use an external device at the same moment. If these people do not have the same goal, their communications with the device will conflict. If the involved persons do consider the external device to be a part of their body because they are directly connected to it, the conflicting action of the other person(s) might be considered as a violation of bodily integrity. People may find themselves restricted in doing with their body whatever they want. It becomes even more critical when there is not only a restriction, but when there is also harm or damage brought to another person including, for example, when programmed applications are erased or modified.

In order to identify to what extent and in which circumstances the right to bodily integrity is affected by these new technologies, we will first give a brief explanation of the right and its international context. This is best explained by using a specific national example of the right, for which I have chosen the Netherlands, since the Dutch Constitution has an entire article (Article 11) dedicated to bodily integrity. Then, the right will be discussed in relation to technological implants. Subsequently, there will be a focus on information carriers, such as RFID and the influence of readable information on the right to bodily integrity.

8.3 The Right to Bodily Integrity

The right to bodily integrity in the Netherlands has a brief, albeit important, history. It was implemented in the Dutch Constitution (*Grondwet*) in 1983 after a discussion that had started in 1976, when the first proposal for a provision

---

[5] Warwick 2002.

containing this right was drawn.[6] The Dutch government was of the opinion that the right to privacy, as laid down in Article 10 of the Dutch Constitution, was sufficient enough to include the specific protection of the human body. However, the parliament did not agree; they unanimously asked for a modification of the Constitution. The government submitted a proposal, in conformity with the request of the parliament, which was accepted without much discussion.[7] Since then, Article 11 of the Dutch Constitution has read as follows:

> Everyone shall have the right to inviolability of his body, without prejudice to restrictions laid down by or pursuant to Act of Parliament.[8]

The scope of the constitutional right is considered to be very wide. The government and the parliament explicitly wanted to cover the right to prevent others from an individuals harming their body (defensive right), as well as the right for an individual to do whatever they wanted with their body (right of self-determination).[9] The government had a duty to make the positive right possible to be executed. There is a duty of care to ensure that a climate arises in which the constitutional right to inviolability of the human body indeed comes to expression.[10] In general a violation of the right is not tolerated by the Dutch courts unless the violation is justified by a right of another individual that should prevail in the particular case. The right is therefore not an absolute right.

It appears that in practice the right is particularly related to criminal law and health law. However, the right also applies to horizontal relationships. This includes, for example, relations between citizens and not only in relations between the government and citizens. The right is meant to provide a protection of the human body, thus to prevent others from infringing the body. Nevertheless, a claim to let other persons perform certain treatments (like euthanasia) cannot be derived from Article 11[11] as the right is a so-called negative one, which facilitates complaining after harm is done. This is in contrast with positive rights that offer a claim towards others to perform certain actions. Negative rights are mostly related to the classical fundamental rights, whereas positive rights are usually socio-economic rights.[12]

As mentioned above, there has been discussion concerning whether a specific right related to bodily integrity was necessary to be contained in the Constitution, as the right might be covered by other provisions concerning privacy rights. For

---

[6] *Kamerstukken II* 1976/77, 13 872, nr. 17.

[7] Koops et al. 2004, p 117.

[8] Koops and Groothuis 2007, p 172.

[9] Koops et al. 2004, p 124.

[10] Zoontjens 2000, p 180.

[11] Koops et al. 2004, p 125.

[12] Classical fundamental rights are, for instance, the right to bodily integrity and the right to life, whereas socio-economic rights are, for instance, the right to work, education, health and housing.

instance, the right to privacy and family life[13] also offers protection against others entering a personal or private sphere. The fact that adding a specific provision is not that self-evident also becomes clear when looking at other countries. Since 1994 Germany has had a similar right, as evidenced by Article 2(2) of its *Grundgesetz*.[14] Also Canada (1982) provides for a constitutional protection of bodily integrity.[15] However, in most countries the right is not separated and reference is made instead to other constitutional rights such as the right to privacy, the right to life and, even more often, to human dignity.[16] The relation between bodily integrity and human dignity becomes very clear in cases such as, for example, xenotransplantation, where organs of animals are implanted to replace a human organ.

The reason to treat the body with dignity or integrity can be based on the person as being an instance of human life or of the human species: 'A being that is human possesses, to the extent that it is human, an essential dignity and identity, because it is *human*.'[17] This implies that, for example, in the case of xenotransplantation— the implantation of an organ of an animal in a human body—results in a human should be treated with dignity, except for the animal part. Fukuyama, however, has stated that,

> [t]he kind of moral autonomy that has traditionally been said to give us dignity is the freedom to accept or reject moral rules that come from sources higher than ourselves, and not the freedom to make up those rules in the first place.[18]

And that '[h]uman beings are free to shape their own behaviour because they are cultural animals capable of self-modification.'[19] If xenotransplantation should imply rejecting a moral rule, namely that animals and humans should not be combined, Fukuyama's argument defends this because this rejection of dignity exists. In contrast with the first argument then, the human being as a whole, including the transplanted organ, would be treated with dignity. From this discussion the difficulty of human dignity as part of bodily integrity, or, in opposite form, human dignity as an overarching concept, under which bodily integrity exists, becomes clear. The concept is closely related to moral considerations, whereas the right to bodily integrity is meant to be a concretisation of morality. When discussing bodily integrity by using the frame of dignity, it appears to be difficult to exclude moral arguments. Dignity is a basic moral concept, so discussions where dignity is at stake will almost automatically focus on moral arguments. Morality is inextricably bound up with reasoning

---

[13] Laid down in Article 10 of the Dutch Constitution.

[14] Article 2 [Personal freedoms] (2) Every person shall have the right to life and physical integrity. Freedom of the person shall be inviolable. These rights may be interfered with only pursuant to a law. Text (official translation) available at: http://www.bundestag.de/htdocs_e/parliament/function/legal/germanbasiclaw.pdf.

[15] De Hert et al. 2007, p 275.

[16] De Hert et al. 2007, p 275.

[17] Reuter 2003, p 57.

[18] Fukuyama 2002, p 124.

[19] Fukuyama 2002, p 128.

people and can therefore not be excluded. These ethical challenges to the acceptability of implants are the result of a natural technology development, diffusion and acceptance process.[20]

In this chapter, the focus is on voluntary implants, which basically implies that criminal law is excluded from the scope of this chapter.[21] Health law can still be applicable, because implants often imply a medical interference. In the following sections I will—for the sake of clarity—use the term 'right to bodily integrity', to include 'human dignity' and 'privacy', when these rights are interconnected or are similar in meaning. The crucial point of discussion concerning the scope of the right to bodily integrity may appear at the border when treatment and enhancement cannot be categorically separated.

8.4 Biotechnological Implants

While this book is focusing on ICT implants, the legal discussion around enhancement and bodily integrity has to date predominantly centred on 'biotechnological' implants, i.e. human organs, genes or cells that can be implanted or injected. Clearly much of this discussion is relevant and transferable to the area of ICT implants, and so this section gives an overview of biotechnological implants in relation to bodily integrity.

Biotechnological implants may violate bodily integrity. According to the legal interpretation of constitutional rights as applied by Canadian courts, the closer something can be tied to the individual, the higher the expectation of privacy and the protection of the body.[22] Implantation and transplantation are often considered as the most invasive to bodily integrity of the individual. Implanting something is the closest one can get to the individual. Because it is considered to be so invasive, consent of the patient, as a requirement for (certain) medical treatments, shall be very well explained in a legal sense. Without clarity on the scope of the term 'consent' it is impossible to judge whether this consent was given in a particular case. Basically, most biotechnological implants have a medical function. For instance, kidney transplants and heart transplants are quite common as treatments and offer patients good opportunities to live longer. Since these treatments require, as mentioned, the consent of the patient, the infringement of the right to bodily integrity is covered by law; personal decision making about bodily integrity must be respected.[23] However, the exact requirement is *informed* consent. On the scope

---

[20] Marshall 1999, p 86.

[21] In a criminal law context, criminals may be forced to have an implant which enables tracking and monitoring at a distance. This facilitates monitoring of convicted persons on probation or with home parole.

[22] Marshall 1999.

[23] Hartman 2007, p 69.

of this term, discussion is possible. It appears to be difficult to define 'informed consent' properly, in particular when patients have to agree on a certain treatment that is technologically not understandable for a layman.

With respect to bodily integrity, more problematic is what happens if biotechnological implants are not implanted because of a medical treatment, but because someone wants to have another organ, or when someone wants to have cells or genes of another person injected into her body.[24] In this situation, the medical purpose is absent; the aim is to change or alter the body or to enhance certain properties of a person. In these cases there is consent of the person who asks for the implant.[25] However, since there is no medical purpose, even though an implant operation in itself certainly is a medical treatment, the applicability of medical law provisions might become unclear.[26] In order to keep things clear, only the open norm of Article 11 of the Dutch Constitution (the right to bodily integrity) will be discussed here. When there is consent, it might be argued that there is no violation of the right. However, as already discussed in this chapter, the right does not cover a claim to let other persons perform a certain treatment. Although there is consent and the right to bodily integrity seems not be harmed, a tension towards the right comes to being. If persons can just change their body with implants or gene injections, questions arise with regard to the value of the right and the boundaries of the human body that is protected by the right to bodily integrity. Responses to these questions will mainly be determined by moral arguments.

As previously discussed, the difficulty is where to draw the line between therapeutic interventions and enhancement: 'What may be therapeutic in one circumstance may be considered an enhancement when used by healthy individuals and adapted for other purposes.'[27] The distinction may be made between replacements (filling in the missing part) and designs for specialised or enhanced functions.[28] However, the impact of an artificial organ or a xenotransplant goes much further than just mere replacement:

> [L]ike all technological objects the replacement is not a neutral adjunct to the body; rather, depending on the context in which it is used, a new subjectivity may be created for the user, and new meanings of embodiment may be created.[29]

Here, the tension towards the right to bodily integrity crystallises. The way in which a person experiences her body might be changing. The question is whether

---

[24] This situation is future theoretical. However, there are cases where this has already happened, even though the treatments are prohibited by law.

[25] I excluded people who are unconscious or incompetent to give their consent, since they are usually legally represented by someone who is allowed to give consent for them. This is, however, another discussion which falls out of the scope of this contribution.

[26] Basically, these laws do apply, but most treatments like those mentioned are not allowed to be performed, so they take place in secret.

[27] Hogle 2005, p 697.

[28] Hogle 2005, pp 706–707.

[29] Hogle 2005, p 707.

this affects the integrity of the body. Is the body, when experienced in a new way, still the body one had before the biotechnological implant was inserted? And if not, does this mean that bodily integrity does not exist anymore for this person or only partly so? Absence of bodily integrity as a result of an implant appears to be very strange and will, likely, not be the case. But still, there seems to be a difference with other changes to the body. This is because the bodily perception is at stake in contrast with other cases where the body can be seen as an object. A more instrumentalistic view will lead to different outcomes to the questions and more or less reject the moral issues that arise. However, it is questionable to what extent an instrumentalistic view can be upheld in a biotechnological debate.

Changes to the human body are certainly not new. For centuries, tattoos and piercings have been commonly accepted within a number of cultures as permanent alterations. What then exactly is the difference between a tattoo and an implant that makes bodily integrity be at stake with implants, but not with tattoos? Or does the same concern count for tattoos, but are we less aware of that? The most obvious distinction between tattoos and technological implants is that implants become part of the inner body, whereas tattoos are put onto the skin outside the body. The question is whether this really makes a difference. What, for instance, if someone wants to have an extra limb? It would be an adaptation of the body, visible at the outside. The aim, beauty or belief, can be the same for this person as the aim of a tattoo.

It is very difficult to identify exactly what makes biotechnological implants, when used for purposes other than medical, different and why they put the right to bodily integrity under a certain tension. One thing that certainly makes biotechnological implants special is that the implants have a 'living interaction' with the rest of the body. The value of the right might be at stake when there are no limits to alter the body with parts that are also alive and become dependent on the connection to your body to continue existence. On the contrary, it may be argued that the right becomes even more valuable, since parts that are added to the body also need to be treated with bodily integrity. A violation of an added part will likely cause serious damage to the 'original' body because of the direct connection. The tissue of the biotechnological implant and the body will attach to each other, implying that it becomes more difficult to remove the implant after a while.

Here, reference is made to today's liberal society. Indeed, society and social values are subject to change, but not everything will be acceptable. However, given the idea that social and moral values are slightly changing over time and that with these changes step-by-step more practices become acceptable, at some point in time the situation will occur that the right to bodily integrity is valueless since it does not protect any dignity anymore, at least not in the sense that it does nowadays. It can protect dignity in the sense that individuals have the right to modify or enhance their body because they are human beings with moral autonomy.[30] The body, as a living matter, can be changed or enhanced as much as the individual

---

[30] Fukuyama 2002, p 124.

wants, implying that there no longer exists an unequivocal notion of what the human body exactly is. Without this common concept, the right as it is applied nowadays cannot be made concrete in practice. The uniform explanation on which everybody would agree that this should be protected with bodily integrity is lacking.

Even when an unequivocal notion of what the human body is remains—even, when it is mutilated or enhanced—the approach towards the way in which the human body should be treated is changing. Treatments or surgeries that unarguably violate the integrity of the human body, albeit with consent of the patient, have become commonly accepted. This can be elucidated from two different perspectives. One is that absence of a medical need for a treatment in combination with the patient's consent does not imply a violation of the right to bodily integrity. In essence, the violation is 'repaired' by the voluntary character of the treatment. In this approach the right to bodily integrity still exists, but the focus for the application of the right lies on consent. The other perspective to explain the general acceptance of treatments and surgeries is related to the notion of what the body exactly is. From this perspective, the body is the natural body including all possible alterations, mutilations or additions. Because of the fact that all supplementary parts are considered to be part of the body, and therefore are considered to be protected by the right to bodily integrity, the treatments needed for changing the body should also be respected.

To conclude, with regard to biotechnological implants, it can be argued that these implants are usually seen as 'real body parts'. Because of the nature of these implants it is arguably the case more often than with other types of implants (see below). When taking the connection and living interaction with the body into account, the right to bodily integrity can be subject to change in several ways. Either the right is loosing its value, since there is no common notion remaining of what the body exactly is and what exactly should be protected with the right, or the right is extending its value by also protecting the implants. Another approach toward the right of bodily integrity is to differentiate between a focus on the patient's consent and a focus on the notion of the body when examining the applicability of the right.

8.5 Non-Living Implants

In the previous section, the debate concerning implants in relation to bodily integrity was outlined from the perspective of biotechnological implants. This debate can to a large extend be transferred to the context of other types of implants. The major difference is that with biotechnological implants there is a living interaction between the implant and the human body; with other implants this is not the case. Another distinction that may be made is that concerning non-living implants and the distinction between active and passive implants. Active implants can influence performance of the individual or fulfil certain functionalities, such as giving signals to the body, whereas passive implants only monitor characteristics or simply carry data, such as monitoring glucose levels or having medical data readable on a chip. This section reflects on the specific issues concerning other implants with a specific focus on nanotechnology and information carriers.

8.5.1 Nanotechnological Implants

This section deals with nanotechnological implants. While the term 'nanotechnology' has been increasingly used by governments, industry, the media and others, a universal definition has yet to be agreed upon. A functional starting point is however the National Nanotechnology Initiative[31] definition:

> ...the understanding and control of matter at dimensions of roughly 1–100 nm, where unique phenomena enable novel applications. Encompassing nanoscale science, engineering and technology, nanotechnology involves imaging, measuring, modelling, and manipulating matter at this length scale.

A single nanometer is the equivalent of one billionth of a metre (10^{-9}), while nanoparticles are particles with one or more dimensions within the 1–100 nm size range.

The small size of nanoparticles makes them very suitable for making products that can be implanted or injected. The fast development—in particular miniaturisation—in the field of computer chips gives input for spectacular speculations, for example mind transplants, and gives perspectives on a broad range of practical applications.[32] In contrast with the biotechnological implants, nanotechnology-based implants are not living. These implants can, however, also interact with the body. One such example is the (hypothetical) curing of diabetes by injecting many nanobots into the patient's bloodstream.[33] These nanobots would be able to synthesize insulin, and to secrete it according to the level of glucose they would sense.

In the case of nanotechnological implants there are only two possible justifications: a medical purpose or enhancement. At first glance, the aesthetic purpose cannot be the case, since the implants are not visible on the outside of the body, apart from the question whether they would be visible without a microscope anyway. However, indirectly the purpose can be aesthetic in cases where beauty mechanisms are influenced by the implants. For instance, certain functionalities can be considered as aesthetic.

In the context of this chapter, it is important to make a distinction between monitored and modified bodies. Monitoring bodies with the help of (nano)technological devices usually has a therapeutic purpose. It can enable regulated

---

[31] National Nanotechnology Initiative. See: http://www.nano.gov/nanotech-101/what.

[32] Van Est et al. 2004, p 39.

[33] Martinac and Metelko 2005.

secretion of medicines, as mentioned above, in combination with monitoring by specialists at a distance. However, this distanced monitoring implies data transfer to external devices. Section 8.5.2 will deal with information carriers and communication in more detail. In this subsection, the focus lies on modification for enhancement. This focus does not necessarily mean that there is no communication with other devices, but it excludes devices that carry personal data.

As an example of enhancement with nanotechnological implants, one can think of brain-machine interfaces (BMI), which translate neuronal activity into command signals for artificial actuators.[34,35] The translation is based on algorithms. In the future BMI may help people to steer devices, such as a car, at a distance by simply thinking. One specific application of the technology is to let disabled people drive their wheelchair without requiring the user to perform any physical action.[36] As mentioned above, nanotechnologies enable the development of devices on a small scale, which are more suitable for implanting. However, the use of nanotechnologies is not necessarily required for BMI.

A completely different issue that comes to mind when discussing nanotechnological implants is the possible risk these implants bring:

[S]tudies examining the toxicity of engineered nanomaterials in cell cultures and animals have shown that size, surface area, surface chemistry, solubility and possible shape all play a role in determining the potential for engineered nanomaterials to cause harm.[37]

For instance, the breakdown of a nano-based coating of an implant might probably lead to toxicity. The question is whether the risk and uncertainty of what exactly will happen is acceptable, or whether humans should not be exposed to these risks. At least it is known that some nanomaterials may, under the certain and limited conditions, penetrate intact skin, implying that controlling the movement of these particles is difficult.[38] There is, however, a debate going on whether skin penetration by nanoparticles is really possible and under what conditions this translocation may occur. At least, there seem to be some conditions which can be of help, but there is no certainty at this current moment.[39] It seems that current concerns over potential risks focus on free nanoparticles as opposed to those that are fixed in a matrix.[40]

The exposure of the body to risks and uncertainties with regard to health can also be considered as an infringement of the right to bodily integrity. Posing risks and accepting possible harm is closely related to human dignity. When using Fukuyama's description of why humans have dignity unpredictable risks may violate dignity

---

[34] Lebedev and Nicolelis 2006.

[35] A more technical description of the technology can be found in Wolpaw et al. 2002.

[36] Millán et al. 2004.

[37] Maynard et al. 2006.

[38] Rousse et al. 2007.

[39] For an overview of research on this aspect, see for example: Australia Government 2006.

[40] Maynard et al. 2006.

since the objective choice to change or modify the body is absent. A person who wants to have nanocomputers injected or implanted into her body only states that she wants that, but the desired change or modification may not be exactly defined.

8.5.2 Information Carriers

This subsection deals with implants which are not only able to communicate with external devices, but which are also capable of transferring personal data to these devices, the implications of which will be analysed in Chap. 9. Such transferring is to be considered in a broad sense; so not only 'real' transfer is involved, but also implants which can be read by an external device, without needing to touch the physical body, like for instance RFID,[41] which can be read at a distance. When these devices contain personal data, this implies that these data can be read from a distance without the person, whom they concern, necessarily being aware of that.

The extent to which this 'data reading' is an infringement of the right to bodily integrity depends on the explanation of the right to bodily integrity. Is the essence of the right related to physical violation of the body, or is the focus more on the protection of information about the body?[42] When physical violation of the body is considered to be a necessary factor, data reading does not imply an infringement to the right, because the body is not affected. However, the right to privacy, which is closely related to the right to bodily integrity, can be harmed. This is particularly the case when the person involved has no knowledge of their data being read.

In this respect, human dignity is of importance since this concept usually covers bodily integrity as well as privacy rights. In the case where bodily integrity is considered to protect information about the body, data reading means an infringement of the right. In this case, privacy, or more specific the protection of personal data, can be seen as part of bodily integrity. Basically, the right to privacy is meant to protect the individual from being monitored or should give the individual the right to determine whether information is disclosed and to whom. With respect to bodily integrity, the right can be used to protect information concerning the body from being accessed by others without the consent, or even knowledge, of the individual.

From a privacy perspective, the majority of European citizens and other RFID technology stakeholders have asked for serious consideration with regard to the use of RFID technologies.[43] Privacy considerations are mostly connected to registration of goods and tracking and tracing possibilities, which lead to extensive profiling. Obviously, tracking and tracing becomes ultimately personal when the tracked device is implanted into an individual and not attached to a product which can be

---

[41] RFID stands for Radio Frequency IDentification, as discussed in Chaps. 2 and 3, for example.

[42] Koops et al. 2004, p 182.

[43] Commission of the European Communities 2007, p 18.

left behind or passed on to another person. When the traceability is connected to personal information that is carried by the RFID device and can be read by external devices, the tension towards privacy rights becomes critical (for more discussion on this, see Chap. 9 of this book). Moreover, the use of such technologies can increase the danger for identity theft and endanger consumer anonymity.

Although there certainly is a tension in relation to privacy rights, it can be questioned whether there are specific concerns for bodily integrity. As discussed above, privacy may be seen as a part of bodily integrity when data that can be read from RFID devices are personal data or contain information about the body. However, the fact that these data are carried on implanted devices does not necessarily make a difference in comparison to situations where the data are read from a chip card or passport. The mere fact that personal data are read secretly is an infringement of privacy rights. The only difference might be that it is impossible to remove the implanted RFID devices, while a passport can be left at home. The voluntary implantation of an RFID device or other information carrier requires informed consent. This informed consent should then include proper knowledge about the impossibility to shield your personal data, which are stored on the devices, from being read.

8.6 Other Implications of ICT Implants and Their Potential Legal Implications

The use of ICT implants may have a range of other implications, next to possible infringements, on the right to bodily integrity such as various cultural effects. In this respect, it is striking that these effects can be the result of different ideas. On the one hand, general acceptance and use of implants can influence legal concepts, as mentioned above. However, on the other hand, there is a risk in not participating in technological developments related to ICT implants. For instance, deafness is sometimes considered as a characteristic of a person and not as a handicap. People in the 'deaf-pride community [consider] deafness a cultural identity, not a disability to be cured.'[44] Not everyone will agree on this. As a result, different opinions may have negative consequences, related to discrimination, stigmatisation and exclusion.

Discrimination may occur if people do not participate in the uptake of implant technologies. They may become relatively inferior to others, because they are not enhanced. In this sense, choosing to remain an original human being naturally leads to discrimination. The other way round, discrimination may occur towards people with implants if they remain a minority.

A generalising effect of discrimination may result in stigmatisation of the subject group of people and then (social) exclusion. This exclusion may even go further towards exclusion from certain public services or health care, since people

---

[44] Sandel 2007.

have chosen to change their body with implants or since they have chosen not to have implants.

Other human rights that may be affected by the use of human ICT implants are the right to life, the right to privacy and arguably socio-economic rights, such as the right to education. The right to life, for instance, may be at stake when artificial hearts become more common. Having a heart implant can then prolong life expectations of individuals. But would this have an impact on the right to life of people who do not want to have such an implant? Privacy rights may be affected when implants carry information about the individual which can be transmitted to other devices. The right to privacy will be discussed more elaborately in Chaps. 6 and 9 of this book.

Finally, consider the case where human ICT implants can be used to enhance certain skills, such as memorising or learning things. Would this influence the scope of the right to education for people who refuse to have these implants and, thus, remain dependent on their natural capacities?

It remains difficult to have a clear vision on the implications of ICT implants and which side will become reality, if one. In any case, the use of human ICT implants may challenge legal concepts and put fundamental rights under tension.

8.7 Conclusions

In this chapter, human ICT implants have been discussed in relation to human rights, with a particular focus on the right to bodily integrity. Implanting an ICT device into the human body implies access to that body. Accessing the body may be a violation of the right to bodily integrity. However, it may be the case that the individual himself wants to receive the implant and gives his consent. The important issue is then to determine the scope of the right. Is it only defensive, in the sense that it can withhold someone from accessing the body, or is it also a positive right, which may support a request for having an implant?

At the moment, the right may not be invoked to force someone else to implant a device into your body. This is important, because, as a result, having an implant will usually require a medical indication. The distinction between therapy or cure and enhancement is sometimes difficult to make, so exactly defining what is deemed medical and what is not is also a difficult issue.

Enhancement of the human body with ICT implants is at this stage. However, in the near future, applications may become available that will make the discussion of the acceptability of human ICT implants necessary. In this respect, it is important to keep bodily integrity in mind. Also other human rights, such as privacy and anti-discrimination, may be influenced by these developments. Although most applications are still theoretical it is not too difficult to come up with some scenarios. Let these scenarios be a guideline in order to have the human rights debate before the scenarios become reality.

References

Australian Government, Department of Health and Ageing Therapeutic Goods Administration (2006) A review of the scientific literature on the safety of nanoparticulate titanium dioxide or zinc oxide in sunscreens

Commission of the European Communities (2007) Results of the public online consultation on future radio frequency identification technology policy; The RFID Revolution: Your voice on the challenges, opportunities and threats. Brussels, 15.3.2007, SEC (2007) 312, COM(2007) 96 final

De Hert P, Koops EJ, Leenes RE (2007) Conclusions and recommendations. In: Leenes RE, Koops EJ, De Hert P (eds) Constitutional rights and new technologies. T.M.C Asser Press, The Hague, pp 265–286

European Group on Ethics (2005) Ethical aspects of ICT implants in the body: opinion presented to the commission by the European Group on ehics. Brussels, 17 March 2005, MEMO/05/97

Fukuyama F (2002) Our posthuman future; consequences of the biotechnology revolution. Picador, New York

Hartman RG (2007) The face of dignity: principled oversight of biomedical innovation. Santa Clara Law Review 47:55

Hogle LF (2005) Enhancement technologies and the body. Annu Rev Anthropol 34:695–716

Koops EJ, Groothuis M (2007) Constitutional rights and new technologies in The Netherlands. In: Leenes RE, Koops EJ, De Hert P (eds) Constitutional rights and new technologies. T.M.C Asser Press, The Hague, pp 159–197

Koops EJ, Van Schooten H, Prinsen MM (2004) Recht Naar Binnen Kijken. ITeR 70, Sdu Publishers, The Hague

Lebedev M, Nicolelis A (2006) Brain-machine interfaces: past present and future. Trends Neurosci 29(9):536–546

Marshall KP (1999) Has technology introduced new ethical problems? J Bus Ethics 19:81–90

Martinac K, Metelko Z (2005) Nanotechnology and diabetes. Diabetol Croat 34(4):105–110

Maynard AD, Aitken RJ, Butz T, Colvin V, Donaldson K, Oberdörster G, Philbert MA, Ryan J, Seaton A, Stone V, Tinkle SS, Tran L, Walker NJ, Warheit DB (2006) Safe handling of nanotechnology. Nature 444:267–269

Millán JR, Renkens F, Mouriño J, Gerstner W (2004) Brain-actuated interaction. Artif Intell 159:241–259

Reuter L (2003) Modern biotechnology in postmodern times? Kluwer Academic Publishers, Dordrecht

Rouse JG, Yang J, Ryman-Rasmussen JP, Barron AR, Monteiro-Riviere NA (2007) Effects of mechanical flexion on the penetration of fullerene amino acid-derivatized peptide nanoparticles through skin. Nanoletters 7(1):155–160

Sandel MJ (2007) The case against perfection. Harvard University Press, Cambridge

Van Est R, Malsch I, Rip A (2004) Om het kleine te waarderen...Een schets van nanotechnologie: publiek debat, toepassingsgebieden en maatschappelijke aandachtspunten. Rathenau Instituut, The Hague

Warwick K (2002) Identity and privacy issues raised by biomedical implants. IPTS Rep 67:29–33

Wolpaw JR, Birbaumer N, McFarland DJ, Pfurtscheller G, Vaughan TM (2002) Brain-computer interfaces for communication and control. Clin Neurophysiol 113:767–791

Zoontjens PJJ (2000) Artikel 11. In: Koekkoek AK (ed) De Grondwet. W.E.J. Tjeenk-Willink, Deventer

Chapter 9
Implanting Implications: Data Protection Challenges Arising from the Use of Human ICT Implants

Eleni Kosta and Diana M. Bowman

Abstract The increasing commercialisation of human Information and Communication Technology (ICT) implants has generated heated debate over the ethical, legal and social implications of their use. Despite stakeholders calling for greater policy and legal certainty within this area, gaps have already begun to emerge between the commercial reality and the current legal frameworks designed to regulate it. The aim of this chapter is to examine the effectiveness of the European Union current data protection regulatory framework for regulating ICT implants. By focusing on current and future applications of human ICT implants, the research presented here highlights the potential regulatory challenges posed by the applications, and makes a series of recommendations as to how such issues may be best avoided by jurisdictions grappling with similar emerging issues. In doing so, the chapter draws together the notions of innovation, risk and data protection within the context of a broader governance framework.

This chapter is an updated and reworked version of the paper: Kosta Eleni & Bowman Diana, Treating or Tracking? Regulatory Challenges of Nano-Enabled ICT Implants, *Law & Policy,* *2011*, Vol.33(2), pp. 256–275.

E. Kosta (✉)
Interdisciplinary Centre for Law & ICT (ICRI), KU Leuven, Leuven, Belgium
e-mail: eleni.kosta@law.kuleuven.be

D. M. Bowman
Department of Health Management and Policy, School of Public Health and the Risk Science Centre, The University of Michigan, Ann Arbor, MI, USA
e-mail: dibowman@umich.edu

M. N. Gasson et al. (eds.), *Human ICT Implants: Technical, Legal and Ethical Considerations,*
Information Technology and Law Series 23, DOI: 10.1007/978-90-6704-870-5_9,
© T.M.C. Asser Press, The Hague, The Netherlands, and the author(s) 2012

Contents

9.1 Introduction

Information and Communication Technologies (ICT) will play a fundamental role in future advances, including within the fields of health care, biomedical sciences, electronics and optoelectronics and military applications, to name just a few.[1] While these advances promise tremendous societal benefits, as with earlier technological advances, such benefits are likely to be accompanied by a range of potential physical, social and ethical risks.[2]

There can be little doubt that human ICT implants, including those which are designed for therapeutic purposes and/or non-therapeutic ones, will become increasingly more sophisticated in the coming years, and offer greater functionality. This is likely to give rise to implants that have the capacity to store increasing volumes of information pertaining to the individual in which the object is implanted. Such data, depending on its nature and the legislative framework in the given jurisdiction, may be defined as personal data. This will give rise to a number of legal obligations for the storage, processing and use of such data in order to ensure the protection of the data subject.

The aim of this chapter is to examine the legal challenges posed by ICT implantable devices under the current European Union (EU) data protection regulatory framework since difficult questions in regard to its potential application to

---

[1] Anton et al. 2001; Kabene 2010; see also Chaps. 2–6 of this book.

[2] Maynard 2006; Weckert 2007.

human ICT implants are apparent. The EU data protection legislation ensures a high level of protection for European citizens with regard to the processing of their data and at the same time foresees a number of strict obligations for those responsible for processing personal data (the data controller[3]). The deployment of ICT human implants in Europe may invoke the application of the European data protection legal framework, when used on European citizens. Even when human ICT implants are manufactured outside the EU, their potential to collect and process personal information upon their use and deployment in Europe will most likely trigger the application of the relevant European legal framework and will incur a number of obligations for the data controllers. By focusing on current and future applications of human ICT implants, this chapter commences by introducing human ICT implants and the broader regulatory challenges presented by them. The focus of this chapter is on the Data Protection Directive and the data protection challenges arising in the context of the EU legal framework and proposed by technological developments, as explained above. The potential application of the ePrivacy Directive to human ICT implants, which contains specific rules on privacy and data protection in the field of electronic communications, is also examined in view of the recent amendment of scope of the Directive relating to radio frequency identification (RFID) devices. The chapter concludes by drawing together the notions of innovation, risk and data protection within the context of a broader governance framework.

9.2 Current and Future Human ICT Implants

The first generation of human ICT implants have, as detailed in Chaps. 2–4 primarily been medical implants. There have, however, been a number of implants that have included subcutaneous RFID devices, designed primarily for personal identification and tracking purposes (see also Chap. 3 of this book).[4] These early implants have been increasingly deployed for commercial purposes including identification, for example, the Baja Beach Club—with night clubs in Barcelona and Rotterdam—to enable VIP clientele to gain access to their venues and carry out payments[5] (see also Chaps. 2 and 3), or for identifying potential kidnap victims, an application reportedly to have been used in Mexico and Russia.[6] While these implants have been traditionally the size of a grain of rice, with dimensions

---

[3] A 'data controller' is defined in Article 2(d) of the Data Protection Directive as 'the natural or legal person, public authority, agency or any other body which alone or jointly with others determines the purposes and means of the processing of personal data; where the purposes and means of processing are determined by national or Community laws or regulations, the controller or the specific criteria for his nomination may be designated by national or Community law'.

[4] Weber 2006; Rotter et al. 2008.

[5] Elliot 2006.

[6] Gosset 2004; Gamboa 2007; McHugh 2004.

of approximately 2 mm by 12 mm,[7] Hitachi advanced the field with the commercialisation of the Mu-chip (or μ-chip), which measure 0.5 by 0.5 mm.[8] While these chips are currently being utilised for applications such as inventory control and security and not for human implant purposes,[9] their development is suggestive of future potential developments in the context of human ICT implants.

Somewhat more controversial due to the potential privacy issues involved has been the development of passive RFID microchips which contain unique numeric identifiers which can then be linked back to a large database containing, for example, personal data such as a person's name, their health, financial or security-related information. By way of example, the US-based VeriChip Corporation[10] has played a leading role in transferring this technology from the agricultural sector to human applications,[11] having developed a human implantable RFID system approximately the size of a grain of rice (see also Chap. 3).[12] The VeriMed patient identification system has been assessed as a 'medical device' by the United State's (US) Food and Drug Administration (FDA), and evaluated on the basis of its therapeutic efficacy pursuant to the comprehensive pre-market regulatory framework established by the *Medical Device Amendments of 1976.*[13] Approval for the marketing of the system as a 'medical device' in the US was granted by the FDA in October 2004.[14] These implantable passive devices are also utilised for identification purposes, with each chip containing,

> a unique identification number that emergency personnel, equipped with an appropriate radio frequency identification (RFID) scanner may scan to immediately identify you and access your personal health information—facilitating appropriate treatment without delay.[15]

According to the manufacturer's marketing material, the devices offer a superior alternative to other identification products as they 'cannot be lost, stolen, or misplaced like other forms of identification'. While it is unclear at this time as to how successful the device has been, it has been reported that within 2 years of the product's approval at least 200 people had been implanted with the device.[16]

---

[7] Wolinsky 2007.

[8] Gamboa 2007.

[9] Hitachi 2007.

[10] The Verichip Corporation is now called PositiveID: http://www.positiveidcorp.com. Accessed 19 August 2011.

[11] Wolinksky 2006.

[12] VeriChip 2007a.

[13] VeriChip 2007a.

[14] VeriChip 2007a.

[15] VeriChip 2006, p 1.

[16] Wolinksy 2006.

The current company R&D focus is on the development of a 'self-contained implantable bio-sensing device included in an RFID microchip' for the purpose of real-time monitoring of blood glucose levels, which would thereby eliminate the ongoing need for invasive testing.[17]

9.3 Emerging Regulatory Challenges for Data Protection in the Context of Human ICT Implants: the Vigilance of the European Legal Framework

Human ICT implants may in various instances contain information about the individual depending on their function. As noted in Chaps. 2–4 of this book, current implants can be used for restoration, enhancement or diagnosis purposes, for identification and authentication etc. In some of these cases the implant functions as an information carrier, bearing, for instance, information about the health condition of the person.[18] In contrast, other types of implants may, however, contain a unique number; this number may be then linked back to specific information about the individual, which may, for example, be stored in a database.

Defining personal data in the European Union in the context of human ICT implants

The Data Protection Directive[19] applies to human ICT implants when they entail processing of personal data. The Directive was articulated in such a way as to ensure that it is technology neutral[20]; this means that it applies equally to all technologies, regardless of what technologies may emerge in the short to medium term. In this respect, the Directive may not be outpaced as quickly as some other regulatory instruments. However, the implementation of the data protection rules and principles included in the Directive need continuous monitoring in order to ensure that they are interpreted and enforced in the 'best lawful, ethically admissible and socially and politically acceptable way'.[21]

> Pursuant to Article 2(a) of the Data Protection Directive, personal data means,
> any information relating to an identified or identifiable natural person ('data subject'); an identifiable person is one who can be identified, directly or indirectly, in particular by reference to an identification number or to one or more factors specific to her physical, physiological, mental, economic, cultural or social identity.

---

[17] VeriChip 2007b.

[18] Halperin et al. 2008.

[19] Directive 95/46/EC of the European Parliament and of the Council of 24 October 1995 on the protection of individuals with regard to the processing of personal data and on the free movement of such data, hereinafter the 'data protection directive', O.J. L 281, 23.11.1995, pp 31–50.

[20] European Commission 2009.

[21] European Commission 2009.

Accordingly, for the Directive to apply, it must be first determined whether the processed data relate to a natural person, and second, whether the data relate to an individual who is identified or identifiable.[22]

When information about an individual, such as, for example, their name, age and/or medical condition, is directly stored on the human ICT implant, the information will be prime facie personal data. The situation may not, however, be so clear-cut when the information is not, for example, directly linked to a natural person but a natural person may, however, be eventually 'identifiable'. This can be the case, for instance, when the information can be reasonably associated to an individual by linking the number of the implant to a back-end database, which contains information about the medical condition of the individual. Recital 26 of the Data Protection Directive provides some additional guidance to practitioners and other relevant parties in regard to how the notion of identifiability should be interpreted. Pursuant to the text of Recital 26, 'account should be taken of all the means likely reasonably to be used either by the controller or by any other person to identify the said person' in order to conclude if we are talking about an identifiable natural person.

Let us now apply this to the aforementioned example of the information that can be reasonably associated to an individual by linking the number of the implant to a back-end database, which contains information about the medical condition of the individual. Pursuant to Recital 26 it is now possible to deem that the information will be considered as personal data. That being said, it is important to note that European Member States have interpreted 'identifiability' in a number of different ways, and have subsequently incorporated their own interpretation of the term into their national legislative instruments.

In view of the need for clarification of the notion of 'personal data', the Working Party on the Protection of Individuals with regard to the Processing of Personal Data, as established under Article 29 of the Data Protection Directive (the so-called 'Article 29 Working Party',) adopted an opinion on the concept of personal data.[23] Pursuant to this Opinion, the Article 29 Working Party has stated that the technological state of the art at the time of the processing, as well as the future technological possibilities during the period for which the data will be processed, have to be considered in deciding on the identifiability of a person. More specifically, the Article 29 Working Party expressed the opinion that,

> if [the data] are intended to be kept for 10 years, the controller should consider the possibility of identification that may occur also in the ninth year of their lifetime, and which make them personal data at that moment.[24]

One can understand the reasoning behind this statement which requires that data controllers should consider the 'possibility of identification' throughout the

---

[22] Pitkänen and Niemelä 2007.

[23] Article 29 WP136 2007.

[24] Article 29 WP126 2007, p 15.

storage period of the data. However, there are significant arguments against the feasibility of such an approach. Since technologies such as ICT are emerging and maturing with great speed, it would appear that in many cases it will be difficult—if not impossible—for the controller to foresee the means that might be used within the short to medium term of the storage period of the data. Moreover, trying to foresee the possibility for identification in the longer term, such as in a 10-year period, would in some cases appear to be simply unrealistic for a controller.

9.4 Processing of Personal Data and the Regulatory Challenges with Regard to Human ICT Implants

9.4.1 Defining the Controller and the Processor in a Human ICT Implant System

The EU data protection legislation distinguishes between two principal actors—besides the data subject—with regard to the processing of personal data: the data controller and the data processor.

Pursuant to Article 2(d) of the Data Protection Directive, the controller is defined as a person (natural or legal) which alone or jointly with others 'determines the purposes and means of the processing of personal data'. In contrast, the processor is a third party who simply processes personal data on behalf of the data controller without controlling the contents or use of the data (Article 2(e) of the Data Protection Directive). This distinction is of great importance in the processing of personal data that are directly stored in human ICT implants or can be linked to an individual. The data controller (and not the data processor) is the party who will carry the obligations described in the Data Protection Directive and is the one who is required to define the details of the data processing. When personal data is stored on a human ICT implant, the controller will be the party who has chosen that the implant will be used as a medium to store personal data and has decided which type of data will be stored on the tag.[25] As a rule of thumb, it can be said that the data controller is liable for violations of the data protection legislation, while the role of the data processor is reduced.[26] In some cases, there may be multiple controllers for the same data set or multiple processors.[27] It is, however, the entity which has the final decision-making power with regard to the processing at issue (the type of information being processed, access to the information, transmission of data, ...) that should be labelled the controller. Once an entity

---

[25] See Zwenne and Schermer 2005, where they applied the same reasoning to RFID tags.
[26] Kuner 2007.
[27] Kuner 2007.

becomes sufficiently involved in the determination of the purposes and means of the processing, it will be considered a (joint) controller.[28]

9.4.2 Legitimate Data Processing of Data Stored on or Transmitted Via Human ICT Implants

Human ICT implants, whether used for therapeutic or non-therapeutic purposes, currently involve the processing of personal data in two ways:

1. personal data may be directly stored on the implant in the form of, for example, information about the medical condition of a patient, or
2. by combining information that is available on the implant, like a unique identifier, with data that are stored somewhere else, such as in a database.[29]

However, in the ubiquitous networked societies of the future, human ICT implants may be used as the means to easily identify an individual. This has the potential to enable the tracking and tracing of the implant, and consequently of the individual, at all times.

The processing of personal data is allowed and thus legitimate only under the grounds foreseen in Article 7 of the Data Protection Directive or under the conditions set out in Article 8 of the Directive, when processing of sensitive data takes place. As such, each time personal data are processed—by means of collection, recording, storage, adaptation, alteration, retrieval, consultation, disclosure, dissemination, etc.—the controller has to verify if the processing falls under one of the criteria for making the data processing legitimate.

The processing of personal data is deemed to be legitimate under the text of the Directive when the data subject has unambiguously given their consent. The 'data subject's consent' is defined under Article 2(h) of the Data Protection Directive so as to include any freely given specific and informed indication of their wishes by which the data subject signifies their agreement to personal data relating to them being processed. The processing is similarly legitimate when it is necessary for the performance of a contract to which the data subject is party or in order to take steps at the request of the data subject for entering into a contract. The processing is also authorised when it is necessary for compliance with an obligation to which the controller is subject. The processing of personal data may also be legitimate when it is necessary to protect the vital interest of the data subject. Finally, the processing of personal data will be considered to be legitimate when it is necessary for purposes of the legitimate interests pursued by the controller or by a third party or

---

[28] Kuner 2007.
[29] Zwenne and Schermer 2005.

parties to whom the data are disclosed, except where such interests are overridden by the interests or fundamental rights and freedoms of the data subject.[30] The processing of personal data in relation to human ICT implants has to be based on one of the aforementioned grounds and compliant with the principles that are set out in Article 6 of the Data Protection Directive in order for it to be considered to be legitimate.

One basic principle for the processing of personal data is that the data shall be processed fairly and lawfully (Article 6(1)(a) Data Protection Directive). It is crucial that the data are collected for 'specified, explicit and legitimate purpose and not further processed in a way incompatible with those purposes' (Article 6(1)(b) Data Protection Directive). The legitimacy of the purposes will depend on various factors, such as the entity carrying out the processing, the environment in which the processing takes place etc. In the context of medical ICT implants, 'the Convention on Human Rights and Biomedicine provides that test predictive of genetic diseases "may be performed only for health purposes or for scientific research linked to health purposes"'.[31] Furthermore, the data controller shall ensure that the collected data are 'adequate, relevant and not excessive in relation to the purposes for which they are collected and/or further processed' (Article 6(1)(c) Data Protection Directive). The procedure followed for the collection of data shall be transparent for the additional reason that in this way the criteria used for choosing the specific data as appropriate can be easily checked. The data shall be kept in a form, which permits identification of data subjects for no longer than is necessary for the purposes for which the data were collected or for which they are further processed (Article 6(1)(e) data protection directive). The data shall also be 'accurate and, where necessary, kept up to date' (Article 6(1)(d) Data Protection Directive).

9.5 Data Protection Rights of the Individual

Human ICT implants may allow the potential collection of personal data stored on them in a continuous and unnoticed manner. The users, therefore, need to be informed about the identity of the data controller at the time the implant is embedded in their body, so that they are able to exercise their rights relating to the processing of their data.

According to Article 10 of the Data Protection Directive, the data controller must provide the data subject with information relating to the processing of their personal data. Such information includes the identity of the controller and their representative, if any, and the purposes of processing. Additional information may be also necessary, such as the recipients or categories of recipients of the data,

---

[30] Buchta et al. 2005.
[31] European Group on Ethics 2005.

whether replies to the questions of the data subject regarding the processing of their data are mandatory or voluntary, as well as the potential consequences of failure to reply. Furthermore, as set out in Article 10 of the Data Protection Directive, the controller should inform the data subject about the existence of the right of access to and the right to rectify the data, insofar as such information is necessary. As a rule of thumb, it shall be noted that the controller shall provide the data subject with all the aforementioned information before the implantation takes place. This is one way of promoting informed consent.

The EU regulatory framework for data protection grants the data subject certain rights that have to be safeguarded by the data controller. The data subject has the right to be informed whether their personal data are being processed. In positive cases the data subject has, for example, the right to know the purposes of the processing, the categories of data concerned and the recipients to whom the data are disclosed. The information must be given to the subject in an intelligible way. Moreover, as set out in Article 12(a) of the Data Protection Directive, in cases of automatic processing of the data, the data subject is entitled to know the logic involved in this. Article 12(b) also grants the data subject a right to ask for the rectification, erasure or blocking of data, the processing of which does not comply with the provisions of the Directive, in particular because of the incomplete or inaccurate nature of the data. Finally, according to Article 14 of the Data Protection Directive, Member States shall grant the data subject the right to object, on compelling legitimate grounds relating to their particular situation, to the processing of data relating to them.

9.6 Applicability of the ePrivacy Directive to Human ICT Implants, with a Focus on RFID Enabled Implants

9.6.1 Applicability of the ePrivacy Directive to Human ICT Implants

There is now a substantive volume of work that has examined the application of the Data Protection Directive to human ICT implants. However, it is not the only Directive that has the potential to raise difficult questions in regard to its potential application to human ICT implants. Directive 2002/58/EC[32] (hereafter the ePrivacy Directive) regulates specific issues regarding the processing of personal data in the electronic communications sector.

The application of this Directive to human ICT implants—and especially RFID enabled ones the most commonly used technology for such implants (see, for

---

[32] Directive 2002/58/EC of the European Parliament and of the Council concerning the processing of personal data and the protection of privacy in the electronic communications sector (Directive on privacy and electronic communications), O. J. L 201, 12.07.2002, pp 37–47.

example, Chap. 3 of this book)—was explicitly mentioned in the ePrivacy Directive. The ePrivacy Directive contains specific rules for the protection of privacy and the processing of personal data in the electronic communications sector, regulating issues such as confidentiality of communications, processing of traffic and location data and unsolicited communications.

The aim of the ePrivacy Directive is to protect the users of publicly available electronic communications services that are offered via public communications networks regardless of the technologies used, having as ultimate goal the achievement of technology neutrality (Rec. 4 ePrivacy Directive). However, questions arise regarding the applicability of the ePrivacy Directive to several emerging technologies, among which is RFID.[33]

For the ePrivacy Directive to be applicable, three main questions should be answered to the affirmative:

(i) Whether there is an *electronic communications service,*
(ii) whether this service is offered in a *communications network* and
(iii) whether the aforementioned service and network are *public.*

Emerging technologies, such as RFID, are usually transmission systems that permit the conveyance of signals in a wireless way,[34] and RFID is a common enabling technology for human ICT implants, as mentioned above. As such, even if these networks differ from the traditional networks that the legislator had in mind at the time of the adoption of the electronic communications regulatory framework, they shall fall under the existing definition of electronic communications networks.[35] When the RFID application is part of a system that can be considered as a service, then they also qualify as an electronic communications service.[36] In the case of human ICT implants, there are applications enabling; for instance, the transmission of data collected through a medical pacemaker to the patient's doctor over the Internet that would qualify as a 'service'. Therefore, it is interesting to examine the third of the aforementioned requirements for the application of the ePrivacy Directive: whether these networks and services fulfil the requirement of being 'public'.

9.6.2 The Requirement for 'Public' Network and Service

The term *public* in the context of electronic communications services and networks was not defined in the EU framework for electronic communications. The lack of

---

[33] Cuijpers et al. 2007.

[34] Cuijpers and Koops 2008, p 884.

[35] Cuijpers and Koops 2008, p 891. The term electronic communications network is defined in Article 2(a) of the Framework Directive (Directive 2002/21/EC, as amended).

[36] Cuijpers and Koops 2008, p 891. The term electronic communications service is defined in Article 2(c) of the Framework Directive (Directive 2002/21/EC, as amended).

any clear definition as to when a network or a service should be, in practice, considered as public has created difficulties in relation to the interpretation of the relevant provisions of the framework, as well as the transposition of the Directives into the national legislative instruments within each of the Member States. Significant problems arise especially in the context of new and emerging technologies that enable the deployment of new services and networks, such as RFID.

The question of whether an electronic communications network or service is public does not seem to have a simple solution. There is a lack of specific criteria on when an electronic communications service or an electronic communications network is public in the legislation or in the relevant policy documents. In view of these *lacunae*, different and various criteria can be—and have already been—listed by different parties in order to define whether or not a network or service should be considered as public. Such criteria could be quantitative, qualitative or a combination of both. For example, one could take into account whether the service or the network has been explicitly characterised as such by the relevant legislator, whether the service is offered by a provider whose primary activity is the provision of electronic communications services, whether it is the provider's intention to offer the service to anyone who requests it, whether the service or the network is aimed at a specific group of people, what is the breadth of the geographical area which is covered by the network or where the service is offered, whether standardised equipment is used etc.[37] The multitude of criteria that may be used for the specification when a service or a network is public reveal the lack of consistent interpretation of the term throughout Europe. Clarification of the term at the EU level is needed; this would positively contribute to the harmonised interpretation among the Member States and to the achievement of increased feeling of security to both citizens and the industry.

9.6.3 Public Communications Networks Supporting Data Collection and Identification Devices

The scope of application of the ePrivacy Directive was modified via the Citizens' Rights Directive. The new Article 3 of the ePrivacy Directive stipulates that under public communications networks are included public communications networks that support data collection and identification devices:

Article 3—Services concerned
   This Directive shall apply to the processing of personal data in connection with the provision of publicly available electronic communications services in public communications networks in the Community, **including public communications networks supporting data collection and identification devices.**[38] (emphasis added).

---

[37] van der Hof et al. 2006, pp 152–153.
[38] Article 3 ePrivacy Directive.

The amendment of Article 3 was inspired by rigorous debates relating to RFID, which has been high on the agenda of the European Commission in the past few years. The Communication on RFID identified the situation that not all RFID applications are covered by the ePrivacy Directive.[39] Taking this situation into account, the Commission wished to clarify that when RFID devices and applications are deployed in connection with public communications networks the ePrivacy Directive should also apply, in addition to the Data Protection Directive. The Directive clarified that when such devices are connected to publicly available electronic communications networks or make use of electronic communications services as a basic infrastructure, the relevant provisions of the ePrivacy Directive, including those on security, traffic and location data and on confidentiality, should apply.[40]

It is questionable whether this amendment to Article 3 of the ePrivacy Directive ('including public communications networks supporting data collection and identification devices') was therefore needed. The amendment that the Directive applies to public communications networks that support data collection and identification devices seems redundant, as the networks in question are public communications networks, which would anyway fall under the ePrivacy Directive. More light on this debate is shed by Recital 56, which clarifies that when collection and identification devices 'are connected to publicly available electronic communications networks or make use of electronic communications services as a basic infrastructure', then the ePrivacy Directive is applicable. However, the Recital makes reference to 'electronic communications services' and not to publicly available electronic communications services. It remains unclear, whether it was the intention of the legislator to broaden the scope of application of the ePrivacy Directive to cover electronic communications services that are not publicly available when data collection and identification devices make use of such services. However, Article 3 is clear that only 'publicly available electronic communications services' and not electronic communications services in general are covered by the ePrivacy Directive and therefore a broader interpretation based on the wording of Recital 56 should not be promoted.

9.7 Discussion and Conclusion

It is evident that technological innovation will play a significant role in driving industrial innovation and economic growth in developed and developing economies. The continual convergence of these technologies with other, more traditional technological platforms such as ICT, promises a range of new products and applications that will enhance human health and well-being. This includes human ICT implants, especially those that offer a therapeutic benefit. However, at this early stage, greater regulatory and scientific certainty is required. And herein lays

---

[39] European Commission 2007, p 5.

[40] Recital 56 ePrivacy Directive.

the challenge: regulating (potential) unknown risks—in terms of ethical, social and legal risks along side any potential human and/or environmental risks—against the broader public interest, while not compromising the development of a promising and powerful application of technology. The foundations for the acceptance of all human ICT implants must be themselves designed and manufactured to ensure their safety and public acceptance. For this to occur, consumers must be confident that the regulatory frameworks that have been implemented are adequate, and that the regulators have the necessary powers and resources to enforce them.

Such legal obligations are not unique to any specific type of human ICT implants. Yet, there is no specific legislation as of today that regulates human implants, irrespective of the technology underpinning their manufacturing and functioning. And as already discussed there is not specific legislation available that deals with the specific data protection issues that arise with regard to human ICT implants. Accordingly, it would appear that this current oversight, when combined with the continual advances being made possible by, for example, microelectronics and nanotechnologies, may be problematic in ensuring an individual's privacy is protected now and into the future.[41] Potential dual purpose applications not only raises issues in relation to privacy and security, but also with respect to informed consent—especially in relation to young children and the elderly—and questions relating to whether or not the system will or should be regulated as a medical device. While we have specifically focused our attention on the EU, these concerns will be not limited to any one jurisdiction. Indeed, they will touch upon and raise questions in every jurisdiction in which such technology become nascent.

One way to alleviate these concerns would be to amend current regulatory frameworks, such as that reviewed in this chapter in relation to the EU. However, given the inherent difficulties associated with predicting the developmental trajectories of the technologies, as well as the inherent functions and characteristics of future devices, it is arguably too early at this stage to fully articulate the dimensions of any such regulatory framework. Accordingly, one practical way to address potential privacy and security threats and to ensure compliance with the data protection legislation would be to encourage the early development and implementation of so-called 'privacy-enhancing-technology' or 'security and privacy-by-design' within human ICT implant system prior to its widespread employment for therapeutic and non-therapeutic purposes.[42] As noted by the OECD in relation to RFID systems generally,

> the availability and adoption of cost-effective and convenient technical safeguards for privacy protection and security might be key success factors for the successful deployment of the RFID. A number of such technical security and privacy controls are already available. Still, cost and technical complexity may remain an obstacle to their deployment in some application areas.[43]

---

[41] See, for example, European Commission 2007; 2009.

[42] Henning et al. 2004; European Commission 2009.

[43] OECD 2008, p 51.

By encouraging researchers and manufacturers to consider the integration of such systems early in the products development, many of the current fears in relation to data protection and privacy may be averted. Such an approach would appear at this time to provide greater flexibility and enable these systems to be specifically tailored to meet the various challenges associated with different implants as the technology and the way in which it can be employed develops and matures. Moreover, by focusing on engineering based solutions as opposed to regulatory solutions at this time, such solutions may be more readily rolled out and adopted across jurisdictions. This would go some way in creating a uniform level of privacy and security protection across different national states, and may arguably install a greater level of confidence about privacy and security within the public arena. Moreover, such security related activities could be employed and integrated into the implants prior to their widespread adoption. This would have the benefit of minimising any such risks to individuals prior to such threats arising.

While there are clearly engineering challenges associated with the development of such systems at the design stage, overcoming these hurdles would arguably be faster than an alternative regulatory response. Such initiatives and solutions would of course be reinforced by existing regulatory frameworks pertaining to data protection and security within jurisdictions and could be actively encouraged by governments through a range of incentives schemes[44]; these could then be amended on an as needs basis for specific applications if and when such a response is warranted.

References

van der Hof S et al (2006) Openbaarheid in het internettijdperk—De invloed van ICT op juridische concepten van openbaarheid (Sdu Uitgevers, Den Haag), pp 152–153

Anton PS, Silberglitt R, Schneider J (2001) The global technology revolution: bio/nano/material trends and their synergies with information technology by 2015. RAND National Defense Research Institute, Arlington

Buchta A, Dumortier J, Krasemann H (2005) The legal and regulatory framework for PRIME. In: Fischer-Hübner S, Andersson C (ed) D14.0.a: Framework V0. Brussels: Privacy and Identity Management for Europe Project

Commission European (2007) Communication on radio frequency identification (RFID) in Europe: steps towards a policy framework. EC, Brussels

Cuijpers C, Koops BJ (2008) How fragmentation in European law undermines consumer protection: the case of location-based services. Eur Law Rev 6:880–897

Cuijpers C, Roosendaal A, Koops BJ (eds) (2007) D11.5: the legal framework for location based services in Europe, FIDIS (Future of Identity in the Information Society) Project, 12 June 2007

Elliot V (2006) Speed through the check out with just a wave of your arm, The Times, 10 October. http://technology.timesonline.co.uk/tol/news/tech_and_web/personal_tech/article666972.ece

European Commission (2009) Commission recommendation of 12.5.2009 on the implementation of privacy and data protection principles in applications supported by radio-frequency identification. EC, Brussels

---

[44] OECD 2008.

European Group on Ethics and Science in New Technologies (2005) Ethical aspects of ICT implants in the human body. EC, Brussels

Gamboa, JP (2007) Micro RFID chips raise some privacy concerns, The Daily Aztec, 22 February. http://www.thedailyaztec.com/2.7446/micro-rfid-chips-raise-some-privacy-concerns-1.798488

Halperin D, Heydt-Benjamin TS, Fu K, Kohno T, Maisel WH (2008) Security and privacy for implantable medical devices. IEEE Pervasive Comput (special issue on implantable electronics) 7(1):30–39

Henning JE, Peter BL, Bernd S (2004) Privacy enhancing technology concepts for RFID technology scrutinised. RVS Group, Bieldefeld

Hitachi (2007) Hitachi RFID Solutions. http://www.hitachi-eu.com/mu/

Kabene SM (ed) (2010) Healthcare and the effect of technology: developments, challenges and advancements. IGI Global, London

Kuner Christopher (2007) European data protection law—corporate compliance and regulation. Oxford University Press, Oxford

Maynard AD (2006) Nanotechnology: a research strategy for addressing risk. Project on Emerging Nanotechnologies, Washington, DC

McHugh J (2004) A chip in your shoulder: should i get an RFID implant? Slate Magazine, 10 November. http://slate.msn.com/id/2109477/

Organisation for Economic Co-operation and Development (2008) Radio frequency identification (RFID): a focus on information security and privacy. Working Party of Information Security and Privacy, Paris

Pitkänen O and Niemelä M (2007) Privacy and data protection in Emerging RFID-Applications. Paper presented at the EU RFID Forum 2007, RFID Academic Convocation, 14 March. http://aalto-fi.academia.edu/OPitk%C3%A4nen/Papers/494866/Privacy_and_data_protection_in_emerging_RFID-applications

Rotter P, Daskala B, Compaño R (2008) RFID implants: opportunities and challenges for identifying people. IEEE Technol Soc Mag 27(2):24

VeriChip (2006) VeriMed patient identification Delray Beach, Florida

VeriChip (2007a) Company: our RFID tags. www.verichipcorp.com

VeriChip (2007b) News release: VeriChip corporation to unveil plans for self-contained implantable RFID glucose-sensing microchip at Grand Hyatt in New York on December 4, 28 November. http://www.verichipcorp.com/news/1196258532

Weber K (2006) Privacy invasions. EMBO Rep 7:s36–s39

Weckert J (2007) An approach to nanoethics. In: Hodge GA, Bowman DM, Ludlow K (eds) New global regulatory frontiers in regulation: the age of nanotechnology. Edward Elgar, Cheltenham, pp 49–66

Wolinksky H (2006) Tagging products and people. EMBO Rep 7(10):965–968

Zwenne GJ, Schermer B (2005), Privacy en andere juridische aspecten van RFID: unieke identificatie op afstand van producten en personen [Privacy and other legal aspects of RFID: unique identification from a distance of products and people], 's-Gravenhage. Elsevier Juridisch, The Hague. http://www.nvvir.nl/doc/rfid-tekst.pdf

Chapter 10
Cheating with Implants: Implications of the Hidden Information Advantage of Bionic Ears and Eyes

Bert-Jaap Koops and Ronald Leenes

Abstract Medical technology advances rapidly. As of 2009, about 188,000 people worldwide had received cochlear implants, and promising trials have been conducted with retinal and subretinal implants. These devices are designed to (partially) repair deaf and blind people's impairments, allowing them to (re)gain 'normal' sensory perception. These medical devices are ICT-based and consist of a sensor that transforms sensory data (auditory, visual, tactile) into signals that can be processed by the brain. Besides data from the regular sensors, in principle, other data from other sources can also be channeled to the brain through the implant, for example wireless data input from distant locations or even the Internet to prompt the bearer with instructions or information. This can be done without others present being aware of this form of techno-prompting, which might give the bionic person a competitive advantage in, for instance, meetings or negotiations. The medical implants could therefore be used for non-medical purposes somewhere in the future. This chapter discusses the normative implications of this hypothetical form of human enhancement, focusing on aspects that are particularly relevant to this type of enhancement as compared to other existing and well-discussed forms of enhancement. In particular, we discuss information asymmetries, ethical aspects related to human enhancement, and some legal issues where the information advantage of bionic sensory implants could make a difference. Based on this discussion, we highlight questions for further reflection and provide some suggestions for the regulatory response to address the challenges posed by the future of bionic sensory implants.

B.-J. Koops (✉) · R. Leenes
Tilburg Institute for Law, Technology, and Society (TILT), University of Tilburg,
Tilburg, The Netherlands
e-mail: E.J.Koops@uvt.nl

R. Leenes
e-mail: r.e.leenes@uvt.nl

M. N. Gasson et al. (eds.), *Human ICT Implants: Technical, Legal and Ethical Considerations*,
Information Technology and Law Series 23, DOI: 10.1007/978-90-6704-870-5_10,
© T.M.C. ASSER PRESS, The Hague, The Netherlands, and the author(s) 2012

Contents

10.1 Introduction

Picture a group of security guards surrounding heads of states with dark glasses and curly black wires connecting their earpieces to the radios in their pockets. These men in black partially receive their orders from intelligence officers in control rooms that bring together the various intelligence sources surrounding the event. We are all aware of the fact that the men in black have prompters, and perhaps part of the effectiveness of this equipment stems from its visibility, similar to the disciplining effect resulting from the knowledge of being observed.[1] But in any case, the earpieces give their users an information advantage.

Of course anyone can buy such earpieces and be in constant contact with helpers outside the scene. This would potentially provide them with a similar informational advantage over others present. This kind of human enhancement would possibly be frowned upon by others and potentially even be disapproved of.

But what if people could have prompters without others being aware of this kind of enhancement?

Medical technology, as history shows up, advances rapidly. As of 2009, about 188,000 people worldwide had received cochlear implants and promising trials have been conducted with retinal and subretinal implants. These implants are information communication technology (ICT)-based and consist of a sensor that transforms sensory data (auditory, visual, tactile) into signals that can be processed by the human brain. They are meant to (partially) repair deaf and blind people's impairments, allowing them to lead a more 'normal' life. The use of these 'neural prosthetics'[2] could in the future go beyond achieving 'normalcy'.

---

[1] Foucault 1978.

[2] As coined by Merkel et al. 2007, p 485.

Instead of connecting regular sensors (microphone and camera) to the implant to create or restore the signal path between the external world and the brain, other inputs can also be used. Neural prosthetics can 'not only *restore* severed sensory functions, but also *computer-enhance* human capabilities'.[3] For instance, audio streams from distant locations can be directly fed into the cochlea prompting the bearer with instructions or advice, similar to the instructions given to the security guards or presenters on television. The (sub)retinal implant can function as an internal 'head-up display' for visual data provided by remote sources. Since the data in these cases are fed directly to the brain, this can be done without others present being aware of this form of techno-prompting.

The implications of cochlear and (sub)retinal implants and their potential of being connected wirelessly to outside data sources have, as far as we are aware, not yet been discussed in the literature. It is important to analyze these implications with due attention to the specificity of the problems raised by this particular type of human enhancement, for, as Bostrom and Savulescu[4] note, we need a 'contextualized and particularized' approach to addressing the question how to deal with human enhancement, applying a case-by-case analysis.

In this chapter, we will therefore focus on the question: what are the normative implications of neural prosthetics with their potential for knowledge enhancement? This is a hypothetical discussion since the technology does not yet exist to connect cochlear or retinal implants wirelessly with outside data sources. It is nevertheless important to start such a hypothetical and perhaps seemingly unrealistic discussion early on since,

> [i]f there is a single lesson to be learned about the past century of scientific and techno-logical discovery, it may well be that the unimaginable rapidly becomes the commonplace.[5]

This chapter provides an explorative account of normative implications of neural prosthetics without aiming to be systematic or exhaustive. We approach the topic from the angle of ethics and law, indicating general issues, primarily from the outlook of liberal constitutional democracies.

This chapter is structured as follows. We start with a discussion of implant technology state-of-the-art and use some cases (Sect. 10.2). The core of this chapter discusses various normative implications of bionic sensory implants, focusing on those aspects that seem to be particularly relevant to this type of enhancement as compared to existing and well-discussed other forms of enhancement. We distinguish between issues related to information asymmetries (Sect. 10.4), other ethical issues related to human enhancement (Sect. 10.5), and some legal issues (Sect. 10.6). We conclude with pointing out issues that require

---

[3] Merkel et al. 2007, p 143.

[4] Bostrom and Savulescu 2009.

[5] Garland 2004, p 29.

further reflection in the academic and societal debate about sensory implants, and provide some suggestions that may help address the regulatory challenges posed by the future of bionic sensory implants (Sect. 10.7).

10.2 Implant Technology

The idea of brain implants already has a considerable history. Merkel et al.[6] quote José Delgado and his colleagues, who in 1976 remarked that with,

> the increasing sophistication and miniaturization of electronics, it may be possible to compress the necessary circuitry for a small computer into a chip that is implantable subcutaneously. In this way, a new self-contained instrument could be devised, capable of receiving, analyzing, and sending back information to the brain.

Today, these implants are a reality, although still in their infancy: the 'most advanced central neural prostheses today comprise the auditory implant, the visual implant, and the human–computer interface (HCI)'.[7]

Cochlear implants (or bionic ears) are the most mature of these three types. The history of cochlear implants goes back to the late 1950s when André Djourno and Charles Eyriès placed wires on nerves exposed during an operation.[8] In 1972, a single-electrode implant designed by Dr. House and 3 M was the first to be approved for implantation into adults by the United States (US) Food and Drug Administration (FDA). Later implants use up to 23 electrodes.[9] Getting a cochlear implant and learning to hear with it is slightly more challenging than accommodating to Vogon speech by slipping a Babel-fish—a tiny fish that translates all languages into your own—into one's ear.[10] A cochlear implant generally consists of one or more microphones, a speech processor that filters and processes the sounds into signals that can be transmitted through a coil held in place by a magnet behind the external ear to a receiver and stimulator secured in the bone beneath the skin. From there the signals are sent to a spiral of electrodes threaded into the cochlea to stimulate the auditory nerve. The patient will have to learn how to interpret the implant's signals—they have to learn how to hear. The quality of hearing with a cochlear implant is (much) lower than hearing people experience. This is not surprising given that even modern cochlear implants have at most 24 electrodes to replace the 16,000 hair cells that are used for normal hearing.

---

[6] Merkel et al. 2007, p 120.

[7] Merkel, et al. 2007, p 121.

[8] See: http://en.wikipedia.org/wiki/Cochlear_implant.

[9] The Cochlear Nucleus CI500 uses 22 contacts allowing for detecting 161 different frequencies. See http://www.cochlear.com/uk/nucleus-cochlear-implants-0 for details. The Advanced Bionics HiRes 90 K uses 16 contacts. See http://www.advancedbionics.com/CMS/Products/HiRes-90K/ for details.

[10] Adams 1980.

However, the sound quality delivered by a cochlear implant is often good enough that many users do not have to rely on lip reading.[11]

Visual implants (or bionic eye), or visual prosthetics, work in a similar way. They consist of an external (or implantable) imaging system (camera) which acquires and processes visual information. The processed data is then transmitted to the implant wirelessly by the external unit (along with power for the implant). The implant converts the digital data into an analog output which is used to electrically stimulate the visual system. The stimulation can be done anywhere along the optic signal's pathway, hence from retina via optical nerve to visual cortex. A major step was achieved when Giles Brindley[12] implanted an 80-electrode device on the visual cortical surface of a 52-year-old blind woman, allowing her to see 'light' (phosphenes) in 40 locations in her visual field. In 2010, the company Retina Implant reported success with a sub-retinal implant consisting of a 1,500-electrode array. One of their patients, a 45-year-old Finland-based male, made the following statement:

> As I got used to the implant, my vision improved dramatically. I was able to form letters into words, even correcting the spelling of my name. I recognized foreign objects such as a banana and could distinguish between a fork, knife and spoon. Most impressively, I could recognize the outlines of people and differentiate heights and arm movements from 20 feet away.[13]

As with cochlear implants, visual implants have a significant difference in quality of vision compared with normal eyesight, due to the enormous difference in the number of receptors involved in this function; the human retina consists, for example, of an estimated 125 million receptors.

The third type of prosthetic implants are sensory/motor prosthetics. Electrode arrays can be implanted into (median) nerves. The array can be used to pick up signals from the underlying nerves, as well as to pass signals to these nerves. In 2002, scientist Kevin Warwick had an array of 100 electrodes implanted in his arm that allowed him to have a robot arm mimic the actions of his own arm, as well as have his arm respond to external signals.[14]

Finally, and more futuristically, there are direct neural interfaces or brain/computer interfaces (BCI). BCI research is aimed at connecting the nervous system directly to computer systems rather than to devices such as cameras and microphones. Until now, BCI research has focused on recording signals from and providing stimuli to animal brains (including, for example, rats, cats, and monkeys). For instance, Stanley et al.[15] have shown to be able to generate movies of what their cats saw and to reconstruct recognizable scenes and moving objects on

---

[11] See: http://en.wikipedia.org/wiki/Cochlear_implant.

[12] Brindley and Lewin 1968.

[13] http://www.businesswire.com/portal/site/home/permalink/?ndmViewId=news_view&newsId=20100317005294&newsLang=en.

[14] Warwick et al. 2003.

[15] Stanley et al. 1999.

the basis of signals from their visual cortex. Several scholars[16] have developed BCIs that decoded brain activity in owl monkeys and used the devices to reproduce monkey movements in robotic arms. Moving from there,

> though futuristic, downloading of the brain may be also relevant in this context. With the possibility of downloading through modified techniques such as those described by Nicolelis at Duke, information from specific areas of the brain, or the whole brain, can be downloaded into a computer. Once downloaded, the information may be modified–for example, by adding a language capacity. Then the altered material may be uploaded into the same individual's brain. (…) This return downloading may in time prove to be an excellent way to enhance cognitive skills.[17]

All these implants start out with the prospect of having people regain or gain capabilities of 'normal' people. As will be clear from the discussion, the implants involve sophisticated signal processing equipment to transform external signals (visual, auditory, sensory) into stimuli for the nervous system. Provided that the number of electrodes in the various types of implants increase significantly, roads are opened for other applications. Information can be superimposed onto visual information provided by the camera to create augmented-reality-like applications directly on the retina or visual cortex. Think, for instance, of turn-by-turn information provided by a TomTom projected on one's visual system, or meta-information about the object being sighted that is directly fed into the brain along with the visual information. The path from medical implants to human enhancement, then, is relatively smooth: any information source that can be transcoded into signals suitable for the electrodes connected to the nervous system, can be employed for improving sensory perception.

10.3 Ethical and Legal Issues

This section discusses ethical and legal issues of neural sensory prosthetics.[18] We will briefly touch upon general issues that arise in other contexts as well, notably with implants and brain enhancement, but we shall particularly focus on issues that might play out differently, or acquire additional salience, when it comes to sensory implants being fed wirelessly from outside data sources. A key difference with other types of brain enhancement that have been discussed in the literature is that sensory implants enhance by feeding content into the brain, and not by enhancing the brain's capacity for processing content. This raises different types of questions from those triggered by, for example, enhancing brain functionality through psychopharmaceuticals or brain stimulation. First and foremost, ethical questions relating to information asymmetries arise (Sect. 10.4). Second, several issues,

---

[16] Wessberg et al. 2000.

[17] Tancredi 2004, p 102.

[18] We will use the terms neural sensory prosthetics and bionic sensory implants interchangeably.

primarily ethical, relate to the enhancement aspect of bionic implants (Sect. 10.5). Finally, legal issues are discussed relatively briefly, since most of these are not specific to the implants under discussion (Sect. 10.6).

10.4 Ethical Issues Relating to Information Asymmetry

Human decision-makers lack the ability and resources to arrive at optimal solutions to problems they are facing. Instead, due to cognitive and time constraints, they necessarily have to simplify the choices available and arrive at satisfactory rather than optimal solutions.[19] Bounded rationality affects any problem-solving or decision-making. In interactions with other people, for instance in negotiations, additional factors contribute to not reaching satisfactory results. For instance, although people negotiate all the time, most are not trained to successfully do so.[20] Hence, they have difficulty in framing, structuring, and thinking creatively about solutions. There are also structural barriers, such as negative emotions and bad atmosphere and power imbalances.[21] People furthermore are hampered by cognitive barriers, such as loss aversion,[22] the anchoring effect, and overoptimistic overconfidence.[23]

Some of these issues can be overcome by providing human negotiators with help from outside. The Man–Machine Interaction Group at Delft University of Technology[24] is currently developing a 'Pocket Negotiator'; a device that helps individuals in negotiations by 'increasing the user's capacity for exploration of the negotiation space, reducing the cognitive task load, preventing mental errors, and improving win–win outcomes'.[25] This pocket negotiator is envisioned to be a smartphone-like device that can be employed in negotiation settings. The Pocket Negotiator is a form of human enhancement, because it may expand and strengthen individuals' capacity to negotiate and to reach better outcomes. Its use, however, may also be contested because it might unduly shift the information position of the parties involved; its use could be perceived as cheating. The Pocket Negotiator gives its user an advantage, both in terms of available information as well as in terms of cognitive capabilities, over the other party.

Neural sensory prosthetics can provide capabilities similar to the Pocket Negotiator, but then (potentially) invisible and embedded in the human body, thereby hiding the information asymmetry from the other party. A bionic attendant of a cocktail party could, for example, obtain useful information about others present at the party. Her

---

[19] Simon 1947.

[20] Thompson 2005.

[21] Thompson et al. 2006; Moore 2003.

[22] Tversky and Kahneman 1981.

[23] Neale and Bazerman 1991.

[24] See: http://mmi.tudelft.nl/negotiation/index.php/Negotiation.

[25] The 'Pocket Negotiator' project proposal, see http://mmi.tudelft.nl/negotiation/images/2/25/Pocket_negotiator.pdf.

camera could recognize people's faces, pull up their Facebook profiles and project, unnoticeably to these others, their profile information about her visual implant. This would make her appear very socially attentive and cognizant of the people present.

Another use of such implants would be augmented-reality applications similar to those already existing for smartphones. An example is a system that overlays information about houses and their inhabitants over the image shot by the smartphone's camera of buildings in a street. Such information, projected on one's implant, could be useful when talking to a realtor selling the property during a visit to this property. Of course much of this information could have been digested by the buyer prior to visiting the house, but if unexpected things happen—for example, when the realtor suggests visiting another house nearby—the instant provision of information about one's built-in head-up display provides the buyer with an advantage, especially because it is hidden from the other party. Another way of looking at this tilting the information imbalance, is that the buyer restores (or creates) a level playing field. Either way, the method by which the bionic individual improves their information position or power may be hidden from the others.

Especially when such applications function without any apparent input from the bionic individual, these applications change the playing field in negotiations and discussions. The information brought to bear by bionically enhanced individuals easily surpasses what other stakeholders in a particular situation may expect of them. In a yard-house sale, sellers may reckon that occasionally a rare expert may turn up who can recognize a piece of value, but they would not expect visitors to come equipped with, for instance, a camera and automated-recognition software that overlays their vision with auction-result web pages. Such bionic enhancement is not necessarily wrong or a form of cheating, but it at least has the capacity to substantially alter social interaction patterns.

Also, cochlear implants may provide unexpected advantages in everyday situations. In an international context we may assume that some people speak multiple languages. This occasionally provides for awkward situations when such people switch to their native tongue for a private exchange, only to discover that unexpectedly someone else also understands their language. When equipped with a cochlear implant and a Babel-fish-like application, someone could easily turn into a polyglot. Youtube, for instance, already provides simultaneous transcribing audio into text (beta); from there, rough translations in other languages could be provided. Even when the translation is not perfect it may enhance one's foreign language capabilities significantly. When others present in a conversation are not aware of this kind of enhancement on the part of one of the participants, this provides the bionic person with an advantage. Again, this need not be ethically or legally wrong, but it does alter the (expectation of) information balance in social interaction.

Information asymmetry always plays a role in negotiations and other communication environments. Buyers are generally less informed than sellers, which causes market imperfections.[26] Also, people differ in their capacities to process

---

[26] Akerlof 1970.

information and to make choices. Classical economics is based on the assumption of perfect information (and perfectly rational actors). Akerlof and others have shown the effects of imperfect information on the side of consumers; their work stresses that many free-market institutions can be seen as ways of solving or reducing 'lemon problems'[27] and compensating negative effects of information asymmetries. For instance, insider trading is prohibited in most jurisdictions, mortgage advisors have to disclose their ties with banks and insurance companies, and financial institutions have to provide financial information leaflets outlining the hidden costs of their services. In many of these cases the potentially distorted expectations of one or more involved parties is corrected by additional information to create a more level playing field. Withholding information or bringing information to bear without informing the others involved may be considered foul play if this information is essential for the decisions at hand.

In the same way, using neural prosthetics linked up to external information sources might in certain situations be considered cheating or unethical behavior. Consider a human-resource manager equipped with the capability of invisibly pulling up information about an applicant, for instance by having her visual support system use facial recognition to retrieve appropriate Facebook information about the applicant. Not informing the individuals under scrutiny of this capacity might be considered unjust, just like people in many countries have to be informed that they are subject to camera surveillance.

It is important to acknowledge that expectations play an important role in determining whether stealthily bringing additional information to bear upon a situation is unethical, and that expectations change over time. For instance, while some years ago it was considered unfair by many human-resource departments to inspect applicants' Facebook profiles, this practice is nowadays less controversial given that every such department is thought to be doing so.

10.5 Ethical Issues Relating to Human Enhancement

Apart from the ethical implications raised by the problem of information asymmetries of implants, we also face normative issues due to the human enhancement aspect of such applications. As we explained in the introduction, sensory neural prosthetics not only restore or establish hearing or sight of impaired people, but they can also enhance functionality beyond normal sensory perception. In particular, they can be fed by wireless signals that are not in the (human) auditory or visual electro-magnetic spectrums, and thus pick up more information than is possible for people without such sensory neural prosthetics, without the

---

[27] The problem that buyers with imperfect information take a risk in buying a product that might turn out a 'lemon', a faulty product.

information input being (necessarily) recognizable. It is clear, then, that bionic sensory implants are a form of human enhancement. This in itself, however, does not raise ethical issues; as Savulescu and Bostrom[28] note, this is only the case if a morally relevant distinction exists between the enhanced and unenhanced functionality. Besides the issue of information asymmetry, which has been dealt with in the previous section, several other aspects seem to merit discussion as they potentially track morally relevant distinctions. We will follow here the general types of arguments offered in the literature on human enhancement, applying these to the concrete application of neural sensory prosthetics.

Before we do so, we can set aside some *topoi* in the human enhancement debate that are not particularly relevant to our topic. These are the challenges, usually raised against particular forms of enhancement or against enhancement in general, broadly based on particular normative outlooks: it is against human nature, it is playing God, or it risks changing the human species beyond its intrinsic nature. Or to phrase it less metaphysically and more eloquently, there is a concern with enhancement 'not as individual vice but as habit of mind and way of being' that reflects an attempt to change 'our nature to fit the world, rather than the other way around', and hence, the enhancement mindset of attempting to gain control over ourselves might lead to a loss of 'openness to the unbidden'.[29] Such objections may have value in certain (particularly dignitarian) normative outlooks, although perhaps less obviously so in our outlook of a liberal constitutional democracy, but in any case there is nothing specific in these objections for bionic implants in comparison with the wide range of enhancement technologies. We therefore leave these objections to the debate in the general enhancement literature.[30] We will also leave aside health and safety issues (except where they touch upon consent issues, see *infra*), and assume that bionic sensory implants do not raise particular health or safety risks. After all, our paper discusses the use of cochlear and retinal implants for information retrieval, and the health and safety risks of this application do not *prima facie* differ from the risks of using cochlear and retinal implants for their primary, medical function of restoring perception. For the sake of argument, we will assume that these medical devices as such are sufficiently safe.

Now we can discuss the remaining ethical enhancement issues as applicable to bionic sensory implants. We distinguish between three major types of arguments: the therapy/enhancement distinction, arguments on the level of the individual, and arguments on the societal level.

---

[28] Savulescu and Bostrom 2009, p 3.

[29] Sandel 2007, pp 96–97.

[30] See for instance Harris 2007; Sandel 2007; Savulescu and Bostrom 2009.

10.5.1 Therapy Versus Enhancement

The first issue is discussed and contested in all enhancement debates: is a particular application acceptable for medical purposes (therapy, restoring functionality) but not for other purposes (enhancing functionality beyond 'normal')? Obviously, it is not easy to define what is a 'medical condition' and what is 'normal' functioning, and the gray zone between therapy and enhancement shifts in time and place. We take a pragmatic approach to this conceptual problem, noting that cochlear and retinal implants, as they are developed and used today, generally have a therapy function, aiming to restore or establish sensory perception that is absent or impaired, while the use of these implants for information retrieval, as hypothetically discussed in this paper, generally does not have a medical but rather an enhancement function. There is an issue whether deafness should be seen as an impairment or a human characteristic,[31] but pragmatically speaking, most people with a cochlear implant to restore hearing would consider that therapy rather than enhancement.

What should concern us here is not the conceptual question, but the material issue whether there is a morally relevant distinction between using sensory implants for therapy or for enhancement. Enhancement advocates would think this is not the case:

> I wonder how many of those who have ever used binoculars thought they were crossing a moral divide when they did so? How many people thought (or now think) that there is a moral difference between wearing reading glasses and looking through opera glasses? That one is permissible and the other wicked?'.[32]

Although Harris has a point here, his rhetorical gusto obfuscates that there may be a moral difference depending on the use of the glasses: if the opera glasses are used not to watch Bryn Terfel on stage but to spy from a distance on Caroline von Hannover sunbathing on her private yacht, some moral border might well be crossed. Of course, the fact that a technology developed for beneficent purposes might be abused by some for malevolent purposes does not imply that the technology should be prohibited outright[33]; it could, however, imply that the development or use of the technology should be regulated to control its potential negative uses. For bionic sensory implants, it is therefore relevant to ask whether they are used therapeutically or for enhancing information retrieval and, in the latter case, whether this is morally acceptable in its specific context. Particularly relevant may be the factor of the implant and its use being unnoticed, or unnoticeable, which makes a key difference between therapy and enhancement here. People interact on the basis of other people having normal sensory perception, and they should take into account (for example, when discussing sensitive issues a

---

[31] See, for example, the ''deaf embryo selection' debate in Wilkinson 2010, pp 66–68.

[32] Harris 2007, p 20.

[33] Brownsword 2009.

little beyond normal hearing distance) that some people have particularly strong hearing, or exceptional visual memory; they will not, however, expect that people they are interacting with have an invisible source of information that directly feeds into their head. Depending on the context, as we have seen in the previous section, this may make a relevant difference to the situation, which will be morally less acceptable if the use of the information stream amounts to cheating in the particular context. There is therefore some reason to believe that, if implants are used beyond therapy, the uses of the implants may need to be regulated depending on their potential for abuse in certain contexts.

10.5.2 Effects on the Individual

A second issue, or rather complex of issues, is the potential effect of the enhancement on individuals. A central concern in the enhancement debate is authenticity,[34] and attitudes toward enhancement seem to correlate quite strongly with how people perceive the enhancement to affect the authenticity of individual human beings. Enhancement sceptics will argue that artificially enhancing people diminishes their authenticity as human beings; they are no longer true to their 'real' self. In a similar vein, Sandel is worried that our appreciation will vanish off the giftedness of humans: the more human functionality is engineered by enhancement technologies, the less human functioning will be seen as a result of gifts people just happen to have, and this tarnishes the humility that is a key feature of our moral landscape.[35] Enhancement advocates, however, could retort that enhancement can also improve authenticity, e.g. by widening individuals' range of choices to 'be themselves' or enhancing their intellectual capacity for being authentic.

Whichever stance one takes toward authenticity and enhancement, we see no immediate concern in applying bionic sensory implants for information retrieval. The implants themselves can hardly be seen as diminishing people's authenticity as human beings—we do not regard a pacemaker or a hearing aid as making someone less authentic in some sense, and there is no fundamental difference between these and a sensory implant in this respect. Nor would the use of such implants for external information input seem to make someone inauthentic, i.e. not her 'true' self, at least not as long as the information input is functionally equivalent to standard forms of information retrieval through normal sensory perception.

A shift might occur, however, once this channel of information input would become so important that it starts replacing other forms of information input: there might be an eerie quality to someone who depends largely on unnoticeable forms

---

[34] Parens 2009.
[35] Sandel, 2007, pp 85–87.

of perception through implants. This is not a short-term concern—we will be entering the cyborg age once this appears on the horizon. The idea of people, and in particular their brains, being connected wirelessly to external information sources, including the Internet, not only challenges our notions of authenticity but also of autonomy and identity. If we imagine a world of cyborgs wirelessly connected to each other via the Internet,[36] it is clear we may need to rethink our concepts of what constitutes the autonomy of an individual, as the boundaries of individual body and mind seem to blur in such a world. Also, individuals' sense of self will be affected if their brains are connected wirelessly to external information sources into which they can continuously tap in real time; as people perceive a tool in their hands to be an extended part of their body, so they could perceive an external information source immediately connected to their brain as an extended part of their mind. Possibly, this could lead to changes in the brain's functioning; for example, the brain could adapt to the implants by storing less information in long-term memory while creating more capacity for quickly processing or connecting information.

This sounds like—and is—science fiction today. But a Warwickian world of cyborgs is a possible (if distant) future, and hence we should not discard the consequences of such a development offhand. It is here that bionic sensory implants do matter, as they can be seen as an initial step on a possible path toward brain/computer-network-interfaced cyborgs. This implies that the consequences of external information sources feeding into the brain for autonomy and identity will have to be taken into account. The short-term implications do not seem particularly significant: as long as the information retrieval remains relatively low-level, individuals' decision-making capacity and their sense of self will not be significantly affected. The mid-term and possible long-term consequences to autonomy and identity may be larger, however, and we should put these on the agenda for further analysis and discussion.

More immediately relevant and somewhat more concrete than authenticity and autonomy in general, is the related issue of the appreciation of results of the enhancement: if someone achieves something, for example answering a difficult question, by means of the enhancement technology, does this affect the value of the achievement? A difference seems to exist in our valuation of human functioning: we tend to appreciate an achievement less when it is 'brought about only by means 'separate and external' to the person using them, thus alienating the person from what he or she achieves'.[37] But whether this matters, depends on the context. In some cases, invisible information retrieval may be seen as cheating, particularly in the context of games or other competitive events. Of course, this depends on the rules of the game, as some games or sports allow deceiving the opponent or hiding communications among the team. A useful boundary mark in this respect is whether the deceptive use of neural prosthetics makes use of

---

[36] Warwick 2002.

[37] Merkel et al. 2007, pp 341–342.

in-game possibilities or whether it violates the rules of the game. This is similar to the distinction made in virtual-world games in relation to, for example, the theft of a dragon sword: if this is done using in-game possibilities, it is part of the game, but if the sword is appropriated by cracking software code or hacking into someone's game account, it is ethically and possibly legally unacceptable.[38]

Outside of competitive contexts, it also depends on the situation whether invisibly using neural prosthetics would be experienced as somehow 'cheating'. If a Briton's cochlear implant would translate speech from Japanese into English, so that she can understand what is being said, we will value her capacity to understand Japanese less than when she had studied the language for years, in terms of appreciation for her language achievement; but we may also value her ability to interact with Japanese which she otherwise never would have had. If someone answers a difficult question after the answer has been fed into his cochlear implant, we will not appreciate this in the context of a quiz or school exam, but we will appreciate it if we simply need the answer ('what is the antidote for a bite by a red scorpion?'). In other words, it depends on whether the focus of the appreciation is on the process or on the result. In most areas of intellectual performance, what counts will be the results rather than the process of achieving them; only in some areas, notably sports—for example, chess—and art as well as knowledge testing situations, will the 'authorship' (an achievement of authentic human effort) impact our appreciation of the result.[39] If we conduct a thought experiment that cochlear implants have an embedded functionality to seamlessly translate foreign language(s), in other words act like a Babel-fish, would this affect our appreciation for people's language ability? It would, in the sense that we would no longer admire people for speaking a foreign language. At the same time, however, for most practical purposes it would not matter at all, and if people want to train their brain or distinguish themselves by showing their intellectual capacity, instead of learning Japanese they could try to master quantum mechanics or some other, non-programmable, feat. Hence, except for some contexts in which the process of getting at information matters, such as sports and knowledge tests, there is no moral objection to information retrieval through implants from the perspective of performance valuation.

A final factor on the individual level is that enhancement can lead to a higher level of individual responsibility: 'As humility gives way, responsibility expands to daunting proportions. We attribute less to chance and more to choice'.[40] Whether this is the case, and whether it matters, is again an issue of context. We can imagine some situations in which more is expected from someone with a bionic sensory implant, for example when she is in a position to provide crucial information in real time (such as the antidote to a red scorpion bite), but in most cases this will not differ fundamentally from someone in a position with any kind

---

[38] Lastowka and Hunter 2004; Kimppa and Bissett 2005.

[39] Merkel et al. 2007, p 353.

[40] Sandel 2007, p 87.

of external information source, such as a smartphone. It is only on a more general level that increased responsibility of enhanced people seems relevant; enhancement through implants could reinforce a tendency to point to people's own responsibility for their destiny: why should I tell you something you want to know, when you can look it up for yourself? This is an argument of potentially diminishing solidarity,[41] which no longer resides in the effect of enhancement on the individual, but in the social responses to enhancement.

10.5.3 Effects on Society

Here we arrive at the third issue: what are the implications of enhancement through neural sensory prosthetics for society at large? The main issue here is distributive justice. In the enhancement debate, this is typically associated with the question whether the enhancement at issue is an intrinsic or a positional good: does it confer an intrinsic benefit (for example, a longer or healthier life) or a benefit only in comparison to non-enhanced people (such as enhanced height)?

Again the distinction is not sharp: certain characteristics, such as intelligence, are positional in some contexts (for example, in job applications) but intrinsic in other contexts (such as being able to enjoy reading Kant), and even typically positional goods may confer an intrinsic benefit in some context (including, for example, tallness enabling someone stranded on a desert island to pick fruit from high trees).

Enhancement of positional goods raises questions of distributive justice: who has access to them, and who are likely to benefit most? But it should be noted that also enhancement of intrinsic goods triggers such questions: although there may be an intrinsic benefit to such enhancement, if access to the enhancement is unequal, socio-economic inequalities may well be aggravated.[42]

How would the enhancement application of bionic sensory implants relate to this issue of 'distributive justice, disadvantaging effects, and the potential for creating an unenhanced underclass'[43]? That would depend first of all on whether people would start having such implants without medical indication, such as when they have normal hearing or sight but want enhanced sensory input. This is unlikely to happen in the immediate and perhaps even mid-term future—apart from health and safety issues, the implants are serious interventions in the functioning of the brain, which will need time and effort to adjust to the new form of sensory input. If less-invasive alternatives are available that are roughly functionally equivalent including, for example, miniature hearing aids or glasses with augmented-reality functions, it is unlikely that people (apart from the likes of

---

[41] Sandel 2007, p 89.

[42] Overall 2009.

[43] Garland 2004, p 26.

would-be cyborgs Kevin Warwick and Steve Mann) will take an implant without medical indication. (It is also questionable whether neurosurgeons will serve healthy people, but that is another issue.) As long as this is the case, there is no special concern with the use of bionic sensory implants; it boils down to the distributive justice of such implants in general. In countries where access to these—expensive—implants is very uneven through inequalities in medical insurance, the disadvantaging effect of implants will be aggravated and there is cause for concern; if on the other hand all impaired people have sufficient access to implants, there is no distributive justice problem.

Suppose, however, that in the more distant future neural sensory prosthetics have become more mainstream and also start to be attractive for unimpaired people. Then the equality issue depends on how costly the implant will be, and which groups are most likely to start using them, and for what purposes. Whether wireless information retrieval through implants serves as a positional or an intrinsic good depends on the context of use; similar to our discussion above of value appreciation, in competitive contexts (sports, knowledge tests) it would be positional, while in many non-competitive contexts (finding a scorpion-bite anti-dote) it would largely seem intrinsic. That suggests that there is no reason to generically restrict access to (non-medical) implants per se, but rather to regulate the context-specific use(s) of such implants. There is one situation, however, in which the enhancement implant as such may have to be regulated, namely when only relatively few people from privileged groups reasonably have access to them. In that case, the implant technology as such could aggravate existing social, economic, or cultural inequalities, and if it would be unfeasible to redress the imbalance by subsidizing or other facilitating measures for underprivileged groups—which might be too costly for public policy—and it might be more appropriate to restrict access to the implants only for people with medical needs.

Another scenario in the same more distant future is that not a few but very many people start using the implants, to benefit from wireless connections of the brain to external data sources. Then the nature of human interactions could change significantly, perhaps radically, particularly if the external sources would be inter-connected in a Warwickian cyborg network. One could argue, as Leon Kass[44] does for life-extension enhancement, that,

> the cumulative results of aggregated decisions (...) could be highly disruptive and unde-
> sirable, even to the point that many individuals would be *worse off* through most of their lives.

But as Brownsword[45] rightly points out, this is hardly compelling in the absence of any realistic basis to estimate and balance the potential beneficent and disruptive consequences. Technologies continuously change social practices, not seldom radically—as with the printing press, the telegraph, and the mobile phone—but change in itself does not provide a basis for ethical concern. If human

---

[44] Quoted in Brownsword 2009, p 135.
[45] Brownsword 2009.

communication and interaction patterns change through bionic sensory implants when many people desire to use them for enhanced information retrieval, there will be a need for close monitoring of potential negative consequences, but not for outright or generic precaution: 'the right thing to do is to make as many better as we can, not to make no-one better.'[46]

This discussion of arguments at the societal level shows that there may be reasons to regulate the enhancement use of bionic sensory implants, but these reasons point to fine-tuned, responsive rather than generic, command-and-control forms of regulation. A precautionary approach aiming to curb the access to or use of bionic sensory implants for enhancement purposes would amount to overkill and a disproportionate limitation of the benefits these implants could have for individuals. Apart from the principled reasoning, practical arguments also oppose a generically negative regulatory tilt: as with many reproductive or other types of enhancement technologies, absent a global consensus that is very unlikely to emerge, people could easily travel abroad to acquire an implant. These would presumably belong to the privileged classes who are able to afford such 'implant shopping', which is another reason why the distributive justice argument suggests a responsive regulatory approach focusing on compensatory measures that ensure that the enhancement is sufficiently accessible to all, that its use does not violate rights of others, and that the enhancement does not harm the infrastructural conditions of a moral community.[47]

10.6 Legal Issues

Many legal issues are relevant for bionic sensory implants, but few of these seem to be very specific for this particular application. Most issues are equally relevant for implants in general. For example, the right to bodily integrity (see, for example, art. 3 Charter of Fundamental Rights of the EU) (see also Chap. 8 of this book) is a key issue for human implants, and informed consent plays a crucial role in exercising this right.[48] Patients must be given sufficient information, in a language they understand, about the health and safety risks associated with implants. The risks of bionic sensory implants are not yet well known, particularly for the embryonic visual and brain/computer-interface implants.[49] Practitioners and patients will therefore be very cautious to use implants unless there is a serious medical reason; the non-medical or recreational use of bionic implants resides in a more distant future.

---

[46] Brownsword 2009, p 135.

[47] Brownsword 2009; Overall 2009.

[48] Beyleveld and Brownsword 2007.

[49] cf. Merkel et al. 2007, pp 402.

Once that future arrives, other legal questions will have to be addressed that are familiar from the enhancement literature, such as whether medical practitioners may implant bionic sensors without a medical indication, and whether employers (such as the military) may force or nudge their employees to take an implant for non-medical reasons. These questions can be left aside for the purposes of our paper, since they are not specific to the use of bionic sensory implants for information retrieval purposes. Two aspects may be specifically relevant, however. The first is that the use of bionic sensory implants for information retrieval, if used on a structural and longer term basis, may affect the structure and functioning of the brain perhaps in other ways than cochlear or retinal implants for 'normal' hearing or seeing do. The brain is known to be plastic and could therefore adapt to structural changes in information input. This could have unknown side effects that will need to be studied, in order to provide would-be implantees with sufficient information about risks and effects to allow them to form informed consent. Another issue is that with information being input into bionic sensors, the likelihood is going to increase that also unwanted or unexpected information is going to be input. In particular, the prospect of computer viruses infecting bionic implants should be studied carefully. This is less science fiction than it may sound: experiments have already shown the possibility of infecting a human RFID implant with a computer virus.[50] Hence, wirelessly connected information-processing implants must be developed and researched with great caution with particular attention to abuse and malware.

Another cluster of legal issues relates not to implants as such, but to information processing and the associated advantage that this type of human enhancement may carry. Here, typical legal issues in information law will apply, such as consumer protection and data protection (see Chap. 9 of this book), as well as sectoral legislation applying to information in particular contexts, such as employment or education. These are also not very specific to bionic sensory implants; for example, data-protection norms will apply equally to processing personal data (such as, for example, being googled through a wireless Internet connection) on a pair of augmented-reality glasses as to processing personal data in a bionic implant. One area where this type of implant may have some specific thrust, however, is sports. Sports regulations apply different standards to using headphones for athletes to be connected to their coaches; while this is regular practice in for example cycling, it is prohibited in other sports, such as soccer. Naturally, it is also not allowed in brain sports that depend on the athlete's information-processing capacity, such as chess or go.

Should bionic sensory implants become in much wider use in the future, each sports area will have to assess whether they have to adapt their regulations to this development. Somewhat more directly relevant—although the technology is still embryonic—is the case when the prospect of bionic sensors implanted for medical purposes could provide some sort of compensatory advantage. Suppose that at

---

[50] Gasson 2010.

some point retinal implants will allow blind or poor-sighted patients to retrieve a substantial part of their eyesight, while at the same time enhancing some sight-related functionality such as eye-hand co-ordination. In that case, some exceptionally gifted people with implants could aim to participate in regular rather than Paralympic competition, for example in archery or biathlon. The case then would be similar to Oscar Pistorius, the 'blade runner' with two carbon-fiber transtibial (below-the-knee) prostheses who was ruled ineligible by the International Association of Athletics Federations to participate in regular competition, as his prostheses were thought to confer a considerable advantage over athletes without such prostheses. However, the Court of Arbitration for Sport (CAS) overruled the IAAF decision, arguing that there was insufficient evidence of an overall net advantage in Pistorius's case.[51] For neural prosthetics used by impaired athletes, similar cases may arise in the future that will call for a (case-by-case) assessment whether they bring an overall net advantage compared to people without such implants; as the CAS observed in the Pistorius case, this is 'just one of the challenges of twenty-first century life'.[52]

10.7 Conclusion

In this chapter we have discussed a particular kind of human enhancement, a class of prosthetics that enhance human (auditory/visual) sensory capabilities and which, because auditory and visual tracts convey information, also indirectly enhance human cognitive capabilities. These implants derive their value not from enhancing the cognitive apparatus ('computing power'), but from providing more and better input for processing ('content'). External intelligence, for example in the form of smart applications that run on external devices, can tap relevant information directly into the nervous system and hence augment other signals entering this system. While similar effects in terms of bringing relevant information to bear can be achieved with external mounted displays and earpieces, the implants are potentially more disruptive in human relations because their presence and functioning may be hidden from other stakeholders present. The implants could, if visible from the outside, also be mistaken for therapeutic devices, such as hearing aids, and thus solicit a sense of sympathy rather than vigilance. When undetected by other stakeholders, these sensory neural prosthetics might uneven the informational playing field to an extent that it becomes foul play. Neural prosthetics are particularly challenging in situations that are crucially dependent on information positions, such as negotiations, knowledge tests, or sports. This could warrant, for example, notification obligations on the part of the bionic

---

[51] Court of Arbitration for Sport (CAS) 16 May 2008, case 1480 (*Pistorius v. IAAF*), available http://jurisprudence.tas-cas.org/sites/CaseLaw/Shared%20Documents/1480.pdf.
[52] Ibid. at 56.

human or other types of context-specific regulations aimed at compensating for the informational advantage of bionic implants.

As we have shown, the effects of bionic sensory implants go beyond the informational plain. In the longer term, if the implants become more prevalent and if people start to use them as a consistent source of information input, the processes responsible for the information feed to the implants will affect how bionic people think, feel, and behave. For example, if the implants constantly provide information on top of, or replace, information perceived by the bionic individual, this may literally lead to tunnel vision, and selective information input or processing could reinforce cognitive biases in bionic humans as well. At the same time, the broader scope of information sources may also widen people's point of view, and it could counterbalance cognitive biases by alerting people to information that their own sensory perception fails to notice. It all depends on how the information feed and the brain are going to interact.

On a more fundamental level, the long-term scenario also has the potential to significantly affect human autonomy, identity, and authenticity. Implants with a consistent external information feed could well be perceived as an extended part of the human body or human being, and because of the wireless connection of neural prosthetics, this challenges our notion of the boundaries of a human being more than is the case with physical technological extensions of the human body. Bionic sensory implants can be seen as a first step toward a future scenario of Internet-connected, and perhaps mutually interconnected, cyborgs. These potential long-term effects call for reflection on how these implants can and should be developed in the short and middle term.

There is no need, however, to be overly precautionary in applying bionic implants, which already have significant therapeutic benefits and which may have equally interesting non-therapeutic benefits in the future. The fact that they also have potential for malevolent use and possible unknown side effects should not lead us to an overall restrictive regulatory tilt. The vision of bionic human beings will not appeal to all and perhaps be scary for a majority of people today, but it would be moral arrogance for us to draw a line of allowing sensory implants only for purely therapeutic reasons. Indeed, it would be moral arrogance to presume we know what is best for future generations based on our current outlooks.[53] As Brownsword reminds us,

> We should not forget (…) that ethical objections to enhancements are not the whole story; even if an enhancement is morally permissible, it does not follow that we should welcome it; but neither does it follow—and this, I think, is the fundamental point—that we have a right to impede the morally permissibly simply because we do not welcome it.[54]

Rather than command-and-control regulation with a negative tilt, we should therefore closely monitor the development of neural prosthetics and discuss their ethical and legal implications on a timely basis. To prevent cheating with bionic

---

[53] cf. Hanson 2009.
[54] Brownsword 2009, p 152.

implants while fostering their fascinating information potential, we recommend a responsive regulatory approach. This should focus on context-specific compensatory measures that ensure that neural prosthetics are sufficiently accessible to all, that their use does not violate rights of others in specific situations, and that they are in line with the infrastructural conditions of a community of human beings centering on social interaction.

References

Adams D (1980) The hitchhiker's guide to the galaxy. (1st American edn.) Harmony Books, New York

Akerlof GA (1970) The market for 'lemons': quality uncertainty and the market mechanism. Q J Econ 84:353–374

Beyleveld D, Brownsword R (2007) Consent in the law. Hart, Oxford

Bostrom N, Savulescu J (2009) Human enhancement ethics: the state of the debate. In: Savulescu J, Bostrom N (eds) Human enhancement. Oxford University Press, Oxford, pp 1–22

Brindley GS, Lewin WS (1968) The sensations produced by electrical stimulation of the visual cortex. J Physiol 196(2):479–493

Brownsword R (2009) Regulating human enhancement: things can only get better? Law, Innov Technol 1(1):125–152

Foucault M (1978) Surveiller et punir. Naissance de la prison [Discipline and punish. The birth of the prison]. Gallimard, Paris

Garland B (2004) Neuroscience and the law. A report. In: Garland B (ed) Neuroscience and the law. Brain, mind, and the scales of justice. Dana Press, New York, pp 1–47

Gasson MN (2010). Human enhancement: could you become infected with a computer virus? Paper presented at the 2010 IEEE international symposium on technology and society, Wollongong, 7–9 June 2010

Hanson R (2009) Enhancing our truth orientation. In: Savulescu J, Bostrom N (eds) Human enhancement. Oxford University Pess, Oxford, pp 357–372

Harris J (2007) Enhancing evolution: the ethical case for making better people. Princeton University Press, Princeton

Kimppa K, Bissett AK (2005) The ethical significance of cheating in online computer games. Int Rev Inform Ethics 4:31–38

Lastowka G, Hunter D (2004) Virtual crime. N Y Law Sch Law Rev 49:293–316

Merkel R, Boer G, Fegert J, Galert T, Hartmann D, Nuttin B, Rosahl S (2007) Intervening in the brain: changing psyche and society. Springer, Berlin

Moore CW (2003) The mediation process: practical strategies for resolving conflict, vol 3. Jossey-Bass, San Francisco

Neale MA, Bazerman MH (1991) Cognition and rationality in negotiation. The Free Press, New York

Overall C (2009) Life enhancement technologies: the significance of social category membership. In: Savulescu J, Bostrom N (eds) Human enhancement. Oxford University Press, Oxford, pp 327–340

Parens E (2009) Toward a more fruitful debate about enhancement. In: Savulescu J, Bostrom N (eds) Human enhancement. Oxford University Press, Oxford, pp 181–197

Sandel MJ (2007) The case against perfection: ethics in the age of genetic engineering. Belknap Press of Harvard University Press, Cambridge

Savulescu J, Bostrom N (2009) Human enhancement. Oxford University Press, Oxford

Simon HA (1947) Administrative behavior: a study of decision-making processes in administrative organizations, 4th edn. The Free Press, New York

Stanley GB, Li FF, Dan Y (1999) Reconstruction of natural scenes from ensemble responses in the lateral geniculate nucleus. J Neurosci 19(18):8036–8042

Tancredi LR (2004) Neuroscience developments and the law. In: Garland B (ed) Neuroscience and the law. Dana Press, New York, pp 71–113

Thompson L, Nadler J, Robert B, Lount J (2006) Judgmental biases in conflict resolution and how to overcome them. In: Deutsch M, Coleman PT, Marcus EC (eds) The handbook of conflict resolution. Jossey-Bass, San Francisco, pp 243–267

Thompson LL (2005) The mind and heart of the negotiator, 3rd edn. Pearson/Prentice Hall, Upper Saddle River

Tversky A, Kahneman D (1981) The framing of decisions and the psychology of choice. Science 211:453–458

Warwick K (2002) I, cyborg. Century, London

Warwick K, Gasson M, Hutt B, Goodhew I, Kyberd P, Andrews B, Teddy P, Shad A (2003) The application of implant technology for cybernetic systems. Arch Neurol 60(10):1369–1373

Wessberg J, Stambaugh CR, Kralik JD, Beck PD, Laubach M, Chapin JK, Kim J, Biggs SJ, Srinivasan M, Nicolelis M (2000) Real-time prediction of hand trajectory by ensembles of cortical neurons in primates. Nature 408(6810):361–365

Wilkinson S (2010) Choosing tomorrow's children: the ethics of selective reproduction. Oxford University Press, Oxford

Chapter 11
Ethical Implications of ICT Implants

Mireille Hildebrandt and Bernhard Anrig

Abstract This chapter focuses on a variety of ethical implications of ICT implants. We will explain how different ethical implications arise from different types of implants, depending on the context in which they are used. After a first assessment of what is at stake, we will briefly discuss the Opinion 20 of the European Group on Ethics of Science and New Technologies as published in 2005. We will extend the scope of discussion by tracing the ethical implications for democracy and the Rule of Law, considering the use of implants for the repair as well as the enhancement of human capabilities. Finally, we will refer to a set of EU research projects that investigate the relevant ethical implications.

Contents

M. Hildebrandt (✉)
Smart Environments, Data Protection and the Rule of Law, Radboud University Nijmegen, Nijmegen, The Netherlands
e-mail: hildebrandt@frg.eur.nl

M. Hildebrandt
Department of Jurisprudence, Erasmus School of Law, Rotterdam, The Netherlands

M. Hildebrandt
Law Science Technology and Society, Vrije Universiteit Brussels, Brussels, Belgium

B. Anrig
Division of Computer Science, RISIS Research Institute for Security in the Information Society, Bern University of Applied Sciences, Biel/Bienne, Switzerland
e-mail: Bernhard.Anrig@bfh.ch

M. N. Gasson et al. (eds.), *Human ICT Implants: Technical, Legal and Ethical Considerations,*
Information Technology and Law Series 23, DOI: 10.1007/978-90-6704-870-5_11,
© T.M.C. Asser press, The Hague, The Netherlands, and the author(s) 2012

11.1 Introduction

Ethical implications can be assessed from different philosophical perspectives.[1] Mainstream approaches to ethics are mostly utilitarian, deontological or based on virtue ethics. From a utilitarian position one would try to calculate to what extent the consequences of human implants satisfy the prevalent preferences of individual citizens, aggregated in order to calculate which usage would achieve the highest (average) good (defined in terms of individual preferences). From a deontological position one would test to what extent human implants violate moral rules or principles, such as human dignity, bodily integrity, autonomy and self-determination or non-discrimination. These two approaches are mostly involved in what is called 'applied ethics,' and both are based on a rationalist analysis and on some form of methodological individualism. Virtue ethics would investigate to what extent human implants contribute to or obstruct human flourishing in terms of virtues like justice, courage, honesty, prudence, etc.[2] The idea of flourishing has been extended to a 'general flourishing ethics' that would include the flourishing of

---

[1] Ethics is a sub-discipline of philosophy. For an overview of approaches in the field of Computer Ethics see Bynum 2001.

[2] The deontological perspective derives from Kant's ethical position, which implies adherence to moral rules that apply even if the consequences of their application are problematic. The moral imperative overrules other considerations. This perspective is focused on the individual choice of a rational person. Methodological individualism also inspires the consequentialist ethics derived from Bentham's utilitarian position, meaning that choices are based on a rational calculation of costs and benefits. Virtue ethics builds on an Aristotelian understanding of human flourishing and excellence, suggesting that we should find the Golden Mean when developing our character, for example the mean of courage between cowardice and recklessness.

non-human entities as relevant ethical implications.[3] Virtue ethics seems less rationalist than deontological theories, but is still focused on individual agents. Although it may be an interesting exercise to chart the ethical implications from these different positions, which each build on distinct assumptions about what is important in human interaction, we will take a different and more practical approach. In tracing the ethical implications we will focus on the extent to which the usage of specific types of implants affects core tenets of constitutional democracy.[4]

We assume that different types of human implants may have different ethical implications, meaning that the analysis will require an acute awareness of both the specific affordances of the technology under discussion and the context in which it is used. After a first assessment of well-known ethical issues such as privacy and surveillance (Sect. 11.2), we will look into 'Ethical aspects of ICT implants in the human body—Opinion 20 of the European Group on Ethics in Science and New Technologies'[5] (in Sect. 11.3), followed by an analysis of the ethical implications for democracy and the Rule of Law, with an introduction to other European projects work in this area (in Sect. 11.4), followed by concluding remarks (Sect. 11.5).

11.2 A First Assessment

11.2.1 Restoration or Enhancement

> When I hear the phrase 'human-implantable electronics,' I must confess that I feel a bit queasy. It conjures up a more extreme image of pervasive computing than is usually justified. However, my perspective is that of a relatively healthy person in his forties, without any physical handicaps. If my hearing was impaired or my heartbeat arrhythmic, I might be keen to find a remedy—and, at this time, an electronic implant would probably be the way to go. Putting my emotional reaction aside, when I think about the possibilities of implantable technology, it actually begins to sound pretty cool.[6]

This statement clearly shows different aspects which might be problematic in the context of ICT implants in general. On the one hand we have a person without physical handicaps who may—at a first glance—be rather opposed to implants 'augmenting' her own capabilities, perhaps not only restricted to physical ones.[7] On the other hand, when it comes to physical handicaps, the same person might agree to have some parts of her body replaced by ICT implants. A first problem

---

[3] Bynum 2006.

[4] Cf. e.g. Nissenbaum 2004; Hildebrandt and Gutwirth 2008.

[5] Rodotà and Capurro 2005.

[6] Want 2008.

[7] See Chap. 10.

here would be to define what really *is* considered as 'augmented.' How should standard human abilities be defined? Consider the case of vision: What *is* a standard human vision capability? A myopic person might answer quite differently to that question than a hyperopic person or even a blind one. Should and can we define an average vision? Though we will not enter this discussion it is important to keep in mind that it all depends on who has the power of definition here.

11.2.2 Defining Implants

At this point we will understand the term implant as any ICT-enhanced item stably connected to the human body, hence we begin with tiny items, e.g. RFID tags implanted under the skin (cf. the Baja beach club example as discussed below), we continue to larger indivisible items like a hip replacement containing some ICT components for—say—measuring physical data with respect to pressures executed on the hip replacement, and we end with major parts of the body (like arms or legs) being replaced or enhanced by artefacts containing ICT components.[8] Hence for the subject discussed here we look beyond conceptual discussions in the context of RFID tags.[9]

11.2.3 ICT Implants and RFID Tags in General

Radio frequency identification (RFID) tags (see also Chaps. 3 and 5) are tiny devices typically attached to the object, allowing data transfer using radio waves with a reader. Often, RFID tags are passive and information can only be read from it, however, there are also active variants which even have their own power supply. A major difference between ICT implants and RFID tags in general is that we consider implants here as typically not being easily detachable from the person, whereas in general RFID tags can be localised on the surface of things or on the clothes of a person. Another difference is that besides RFID tags many other implants are available.

---

[8] We do not want to restrict the analysis to 'standard' ICT implants which are typically considered to be small or even tiny, like RFID implants. We do not want to exclude entire body parts nor nanosized ICT items from this discussion. Neither do we restrict the discussion to *invisible* or to *intangible* implants. After all, what would visibility mean to a human being augmented by say, X-ray vision, or RFID reading capabilities, etc.?

[9] See for example FIDIS deliverable D12.3 'A Holistic Privacy Framework for RFID Applications' (Fischer-Hübner and Hedbom 2007). In this context, one might be interested to look at the general discussion on how ethical responses to new challenging and revolutionary techniques might be developed (Moor 2005).

The Baja nightclub in Barcelona, Spain proposes to its 'most valued' customers the use of an ICT implant to gain several benefits while they are in the club, e.g. for making payments in a more convenient way, using a scanner that every waiter has available.[10] Clearly, there are numerous additional applications which might be implemented in this club based for example on data captured by the scanners, determining profiles or habits of customers. Hence, even using a simple ICT implant, there are a number of potential applications of data mining and profiling. While we do not know which techniques are actually being applied in this case, even a potential application should raise fears of privacy for the 'user', i.e. the client of the bar. The main aggravation of the situation as compared to other forms of RFID tracking is based on the strong link between the person and the implant which will typically last for a long time. From the point of view of the business that applies profiling, this is very interesting since the unique identifier remains stably attached to the person for a long time, hence it will provide 'good' data without much noise. Within the same use case consider for example the potential for determining interesting clients which always bring their friends to the club; such a client is very valuable and may get some free drinks as a 'reward.' Although this may not be new compared to frequent shopper or loyalty cards from super-markets where the concept of personal discounts is technically feasible, the difference is that the implants have a strong link to one person, whereas loyalty cards may for instance be shared between several persons. This will provide a business with more reliable information to treat persons differently based on their habits, like giving them price reductions, excluding them from entering the bar, etc.

11.2.4 Communication Aspects of ICT Implants Within the Context of Health Care

Although the problematic aspects of implants are not restricted to the communication parts of ICT implants, as Halperin et al. mention in the context of security in pervasive health care,[11] the impacts of the communicative capabilities are prominent. Though their application in health care may be a special case, it makes sense to give this application special attention if we consider that the number of medical ICT implants will grow massively in the near future. Halperin et al. describe three aspects[12]:

1. efficient methods for securely communicating with medical sensors,
2. controlling access to patient data after aggregation into a management plane, and
3. legislative approaches for improving security.

---

[10] See, for example, Chaps. 3 and 5.

[11] Halperin et al. 2008.

[12] Halperin et al. 2008, p 34, citing Venkatasubramanian and Gupta 2007.

We will skip the second and third points here, that refer to standard data aggregation problems and the legal framework, discussed elsewhere;[13] the first point, however, is crucial to the discussion here. Clearly, the problems are somewhat related to the case of RFID as often similar restrictions to processing power, transmission issues, power supply, etc. may occur, especially for ICT implants of small size.

While the example of the client of a nightclub in Barcelona concerns a situation where the person is willing to have a device implanted in order to gain special privileges, in some cases a person may not really have a choice. Consider the case of a pacemaker which is typically not optional if you need one.[14] In the case where the capabilities of such a device are extended with—say—wireless communication,[15] the user is faced with an additional problematic aspect. From a medical point of view, the extension of capabilities could make sense (for example to easily get historical data from the pacemaker), but from the point of view of the 'user' (i.e. the patient) such an extension may not always be necessary or desirable. Yet if only devices with such additional capabilities are available, the user has no choice.[16] A result of all this could be that effective detection of people with a pacemaker is possible using wireless technology, to the extent that the implants used lack effective security technology. Applications are manifold, from detecting people that have pacemakers in critical environments (strong electromagnetic radiation occurring in a specialised industrial context may harm pacemakers), tracing such people on their way through a building (e.g. in a hospital, after treatment), to behavioural profiling (assuming that the pacemakers are distinguishable), etc. While there are many 'good' applications, all the usual problems of profiling will appear, notably those of privacy, undesirable discrimination and a general lack of transparency of the profiles that have been mined and applied.[17]

In the same context we have to consider the discussion on whether an extension of physical capabilities is what those offered such extensions really want. As a typical example, some parts of the deaf community are rather critical with respect to cochlear implants, due to various reasons which include medical issues, but

---

[13] See for example Hildebrandt and Gutwirth 2008 for the implications of data aggregation and profiling, and Chap. 9 for a legal analysis.

[14] Clearly, there are people who disagree with this opinion, for example due to religious views, etc.

[15] Note that such wireless communication is not the future but the state of the art in pacemakers which operate in the 402- to 405-MHz band, with 250 Kbps bandwidth and have a read range up to five metres (Halperin et al. 2008, p 32).

[16] Compare, for instance, the case of the vaccination of small children in their first 3 years. Nowadays, vaccination for only one disease is not possible in Switzerland or very hard to get. This means that as a 'client' you are forced to have the combined vaccination (typically three of them), hence progress diminishes your choices.

[17] C.f Hildebrandt and Gutwirth 2008.

also: '[e]xtreme proponents of this view regard giving a deaf child a cochlear implant or hearing aids as akin to 'correcting' the colour of a black person's skin by making them white.'[18] The *US National Association of the Deaf NAD* writes in a position paper that 'The media often describe deafness in a negative light, portraying deaf and hard of hearing children and adults as handicapped and second-class citizens in need of being "fixed" with cochlear implants.'[19] Some argue that deafness is not merely an individual affliction but has cultural aspects, i.e.: 'The deaf community is a culture. They're much like the culture of the Hispanic community, for example, where parents who are Hispanic, or shall we say deaf, would naturally want to retain their family ties by their common language, their primary language, which is either Spanish or in our case its American Sign Language.'[20]

The effect of an implant as in this example may hence provoke cultural effects which may or may not be accepted by the respective persons and provoke resistance within their communities. The same issues might appear when considering ICT implants which give humans abilities they do not have by nature, creating 'cultures' of 'augmented' human beings and 'countercultures' of 'naturals.' Kevin Warwick and Joel Garreau have written about the ethical implications of such novel divisions in society.[21]

11.2.5 Acceptance of ICT Implants in the Context of Health Care[22]

With respect to future social acceptance, we can again look at results from the field of RFID implants, where some recent studies show that social acceptance is low. A study commissioned by the New Jersey Institute of Technology in 2002 shows that 78.3 % of the responders would not be willing to implant a chip in their body.[23] Further evidence in this direction is provided by a

---

[18] See: "The Anatomy of Prejudice, A blog about real and perceived prejudice," May 22, 2006, available at http://wallsmirrors.blogspot.com/2006/05/deafness-is-not-disability-argumentum.html.

[19] See: "NAD Position Statement on Cochlear Implants (2000)," NAD Cochlear Implant Committee, approved by the NAD Board of Directors on October 6, 2000, available at http://www.nad.org/issues/technology/assistive-listening/cochlear-implants.

[20] See: "The Cochlear Implant Controversy," CBS Sunday Morning 2nd June 1998, available at: http://www.cbsnews.com/stories/1998/06/02/sunday/main10794.shtml.

[21] Warwick 2003, and Garreau 2005.

[22] This section builds partly on work of P. Rotter, R. Compañó and B. Daskala from IPTS, partly also published in Rotter et al. 2008.

[23] Perakslis and Wolk 2006.

study[24] by CapGemini mentioning a minority of people being 'very & some-what willing'.[25]

Basically identification technologies have *per se* a low acceptance rate and even more so RFID implants, due to their nature of being *im*planted: 'Implanting an RFID tag is considered intrusive and "creepy" (although placing the chip with a syringe is not considered surgery) and many people are reluctant to have this done.'[26]

It is worth noting that social acceptance depends on the application field and seems to be highest for lifesaving purposes: 'According to the CapGemini survey, the percentage of people saying 'not at all & somewhat unwilling' and 'very & somewhat willing' for lifesaving was 42 and 44 %, respectively, the same for using biometrics for passenger identification in air travel, which is already in place.'[27] With respect to general ICT implants and considering their potential benefits to the general public especially also in the context of health care, negative scores may decrease if there is no discussion about privacy issues in general (extending the identification issues mentioned above).[28]

Furthermore, ICT implants have access to personal data measured from the body of the person, hence typically data which could not be recorded by some detached item. While restrictive use of such data within well-designed medical applications will probably be advantageous for the patient, any abuse may lead to serious damage. The unauthorised collection of data from such devices, or—going further—active interaction with them in order to change some behaviour, will be even more problematic. As an example consider deep brain stimulation for patients with Parkinson's disease. If such a stimulation device is controllable from outside (e.g. by wireless communication), one can easily imagine how its abuse may provoke dramatic results.

---

[24] CapGemini 'RFID and Consumers—What European Consumers Think About Radio Frequency Identification and the Implications for Business' (2005), available at http://www.ca.capgemini.com/DownloadLibrary/requestfile.asp?ID=450. This study has been conducted in Europe (UK, France, Germany, The Netherlands) in Nov. 2004 by means of an Internet panel. Responses were made by more than 2,000 persons over 18 years. Among the goals of the study were the understanding of consumers' awareness and concerns with regard to RFID, their willingness to purchase RFID-enabled products and the importance of the corresponding potential benefits from RFID-technology.

[25] For more information see Rotter et al. 2008.

[26] Rotter et al. 2008.

[27] Rotter et al. 2008.

[28] In this context, it is interesting to see that ethical questions with respect to implants and enhancements of animals are getting more into the public focus. In Switzerland for example, recently a corresponding report (Ferrari et al. 2010) was published, commissioned by the 'Eidgenössische Ethikkommission für Biotechnologie im Ausserhumanbereich' (the Swiss federal commission on ethics for bio-technology in the non-human context). This report was also discussed in 'Die Technisierung der Tiere' (M. Hofmann) published in a widely read newspaper in Switzerland (NZZ Neue Zürcher Zeitung, 30.12.2010, p 11).

11.2.6 Control Over One's ICT Implants

The *Report on the Surveillance Society*[29] mentions that 'Surveillance society poses ethical and human rights dilemmas that transcend the realm of privacy' and further says that 'ordinary subjects of surveillance, however acknowledgeable, should not be merely expected to have to protect themselves.' Hence, tools and abilities should be given to the 'users' to control their implants and their communication to the external world.[30] Clearly, a problem here is the gap between the control of the device that is technically possible and the actual capabilities of the user to make use of these possibilities.

As for implanted RFID tags, the most obvious problem with respect to privacy is the typical lack of control. If wireless communication is available in the implant, how can the communication be controlled, supervised or monitored by the person? In case of problems: how can the communication be shut down (and afterwards turned on again)? Is there a need for emergency access and how is authorisation done in this case?

11.2.7 Ethical Codes of Conduct

Results and findings on ethical aspects with respect to RFID tags[31] can partially be generalised to the field of ICT implants. In contrast to the case of RFID tags,[32] there is to our knowledge no specific code of ethics or of conduct with respect to ICT implants available yet. However, general ideas from the versions of codes specific to RFID tags may to some extent carry over to ICT implants.

Two well-known codes of conduct are the *ACM Code of Ethics and Professional Conduct* as well as the *British Computer Society Code of Conduct*[33]

---

[29] Wood 2006.

[30] Some limited and specialised use cases may require that the user cannot fully control the implant and its communication, taking note of the usual exceptions of Articles 6, 7 and especially 8 of the Data Protection Directive. A special example would be the tracking of prisoners using ICT implants, which would clearly require that the prisoner should nòt be able to change the functionality of the implant. Note that this example would have to be based on the monopoly on legitimate force of the state (German 'Gewaltmonopol'), which requires a number of safeguards as specified in Articles 5 and 6 of the European Convention of Human Rights (apart from the applicability of the right to privacy in Article 8). In fact the idea of using implants to track offenders seems an offence to human dignity.

[31] For a discussion on ethical aspects with respect to RFID tags see Fischer-Hübner and Hedbom 2007.

[32] Sotto 2005, UK RFID Council 2006.

[33] Association for Computing Machinery ACM, ACM Code of Ethics and Professional Conduct, (Adopted by ACM Council 10/16/92), available at http://www.acm.org/constitution/code.html. The British Computer Society, British Computer Society Code of Conduct, 2001, http://www.bcs.org/upload/pdf/conduct.pdf. For a large collection of codes of ethics and conduct see Berleur and Brunnstein 1996.

applicable in the field of computer science.[34] Their content is not focused on RFID tags or on ICT implants, yet the points discussed can be carried over to the present context. These codes aim to function at a metalevel, are rather abstract, containing general statements following the idea that—as mentioned in the ACM Code— 'questions related to ethical conflicts can best be answered by thoughtful consideration of fundamental principles, rather than reliance on detailed regulations.' This issue was already taken into consideration by the so-called *Toronto Resolution*,[35] starting from the observation that the detailed regulation of ethical issues usually makes little sense. This so-called meta-code contains as a first part a preamble containing a text to be added to every code of ethics built on these baselines in order to fix to a certain degree the context of the newly built code. The second part of the meta-code consists of 12 rules to be followed while constructing any new code specific to an application domain. This should allow for coherence between different codes at a more generic level, while also informing the more specific applications.

A fundamental question is whether a single code of ethics will be possible for the general case of ICT implants. We are not sure whether the variety of possible ICT implants will allow the construction of one code of ethics applicable in each case. Either we get a very detailed code that easily becomes outdated with the flux of technological dynamics, or a very general code that does not give help to the users in concrete applications. The framework of the *Toronto Resolution* may be of help here, creating different codes for different types of ICT implants, though what typology is to be used is an open question. An example would be to develop a typology based on the type of application, i.e. nowadays, a standard pacemaker is a rather simple implant in comparison with those used for deep brain stimulation.

A major issue for the effectiveness of ethical codes of conduct is clearly the raising of public awareness. In the case of RFID tags, this is happening following the discussions in the press, yet in the case of ICT implants this process is only starting.

11.2.8 *Privacy Enhancing Methods*

In the present context, a main ingredient for ethically well-founded applications of ICT implants must be the offering of privacy, or rather, privacy enhancing methods. In this section we will not delve into this subject as it has been discussed extensively elsewhere.[36] Weber writes: 'It is important to see that ICT implants and biometrics are not the beginning of a process of evaporating civil rights like privacy but only another brick in the wall. Furthermore, it is vital to understand

---

[34] For an overview of these two codes see Fischer-Hübner and Hedbom 2007, p 34ff.

[35] Fawcett 1994.

[36] See different FIDIS deliverables, for example Cvrček and Matyáš 2007, Kumpošt et al. 2007, Müller and Wohlgemuth 2007, Fischer-Hübner and Hedbom 2007, Sprokkereef and Koops 2009.

that this problem cannot be solved by other and new technology—civil rights protection is a social and political task and not one of engineers.'[37] However, as we have proposed elsewhere, the protection of fundamental values like privacy will require their technological articulation.[38] The problems created by these novel technologies cannot be solved by lawyers or politicians without active involvement of computer scientists and engineers. In that sense Weber may be mistaken and technical experts should become aware of their task in safeguarding privacy.

Among other issues, Halperin et al. consider several aspects of privacy and security goals which gives a good overview of some of the problems arising in this field, even if they not all are focused on ethics[39]:

1. authorisation
2. availability
3. device software and settings
4. device-existence privacy
5. device-type privacy
6. specific-device ID privacy
7. measurement and log privacy
8. bearer privacy
9. data integrity

Consider for example the 'device-existence privacy': knowledge of the existence of an implant might compromise the privacy of the person (cf. above). Authorisation must be done in a privacy protecting way using a secured protocol that is not giving any information—on the implant, the entity doing the authorisation, etc.—to a potential attacker. Even if the device reveals its existence to authorised entities, it should, following 'device-type privacy,' not reveal its type (unless authorised to do so). 'Specific-device ID privacy' means that an attacker must not be able to follow an ICT implant without being authorised to do so. Note that a standard passive RFID tag is a typical item without 'specific-device ID privacy,' since it reveals its unique ID to any reader. Clearly, an implant should not identify (unless authorised to do so) its bearer—hence satisfy the 'bearer privacy.'

11.2.9 Conflicting Issues

Although privacy enhancing methods may provide solutions for some of the problems arising in the daily use of ICT implants, there are several ethical conflicts that are not easily resolvable using those techniques. Their difficulty lies in the

---

[37] Weber 2006, p 14.

[38] Hildebrandt and Koops 2007 and 2010.

[39] Halperin et al. 2008, p 31ff.

conflicting underlying goals depending on the application and even more so on the human being bearing the ICT implant.

In the context of ICT implants, some consider enhanced humans (cyborgs) as not being ethically problematic *per se*. Warwick, for instance, asks rhetorically: 'But should such entities, if indeed they are truly cyborgs, present an ethical problem?'[40] We think that serious ethical conflicts may indeed arise, even in the case of merely restorative ICT implants. Though at this point we are not able to present final solutions or guidelines for resolving these issues we end this section with a list of such conflicts.

1. 'Freedom' against 'health': Consider the situation where medical reasons clearly indicate that you should have a pacemaker implanted. If all devices applicable to your situation which are produced already contain some wireless communication device, you may either accept this fact and have such a device implanted or otherwise face the medical problems induced by not having a pacemaker which—in the worst case—might lead to serious life-threatening problems.

2. 'Supervision' against 'anonymity': In the same context as above, bearers of ICT implants may not be anonymous anymore. In some situations, other devices might identify you through your ICT implant, due to some wireless communication and a respective unique identifier. Supervision in this context may however be a crucial and desired issue, e.g. in a hospital environment, the medical staff might need to supervise a patient with an ICT implant newly implanted in order to guarantee convalescence after the operation. However, using privacy preserving technology, a big step should and can be made here in order to avoid supervision by non-authorised third parties.

3. 'Self-determination' against 'heteronomy,' i.e. from a purely technical perspective, one may not be either able or allowed to change or even read the data contained within an ICT implant, possibly due to good reasons like the complexity, etc. This can be considered as a lack of self-determination (in a restricted sense), not allowing to control all of one's own 'parts of the body'.[41] In some sense, the data contained in the ICT implant can be seen as determining the 'behaviour' of the implant with respect to an externally connecting device. Hence, one loses the capability to control this 'behaviour.' From another point of view we also have the conflict between the ones capable of communicating with their own (or even other's) ICT implants and the ones who cannot due to whatever reasons.

4. 'Security and privacy' against 'safety and effectiveness': This must be regarded first in a technical perspective,[42] but may also have ethical components attached. Typical problems might arise for example when controlling access to

---

[40] Warwick 2003, p 131.

[41] In this context the question arises: what does 'belong' to its own body? To what extent can ICT implants—which may be manufactured and even still owned by some company—be considered as belonging to one's body?

[42] See for example Halperin et al. 2008.

data on ICT implants in emergency situations, where different questions are to be answered by different parties, like: What *is* an emergency situation? Which situation—if some classification is available—gives access to what data? Can a medical doctor not better evaluate which access is needed in a situation and should—or even must—she override the restrictions previously specified by the patient in order to save life?

11.2.10 Neural Implants and Human Enhancement

Probably those implants directly connected to our nervous system will induce the strongest ethical challenges. Consider the special case of deep brain stimulation which is a first step in this direction. Such stimulation may provoke not only the expected effects, but may indeed have side effects such as a change of personality.

Warwick states another fundamental dilemma to be considered here: 'Where the cyborgs represent a powerful ethical dilemma is in the case where an individual's consciousness is modified by the merging of human and machine.'[43] Hence, he considers a dilemma arising within the context of consciousness, or mental 'operations'. Evidently the awareness of the person subjected to this human–machine merging should be our first concern. In relation to the implant itself, Warwick notes that 'the biggest surprise for me during the experiment was that I very quickly regarded the implant as being 'part of the body'. Indeed this feeling appears to be shared by most people who have cochlear implants, or heart pacemakers.'[44] Especially in relation to implants which are in contact with our brain or even influence—say—in some sense our consciousness, many fundamental questions arise.[45]

Consider now the enhancement of the human body in a general manner. As noted above, we can discriminate between an implant's function by describing it as restoration or repair of human capabilities, diagnosis of a biological state of affairs, enhancement of human memory, vision, auditory perception, alertness or other human capabilities, and identification, surveillance or billing.[46] Whereas these distinctions may be well-defined in theory, in practice we see major problems when actually being forced to discriminate between the repair and the enhancement aspect of an ICT implant. Where should the line be drawn? If a blind person gets her ability to see in a way that her vision is far more powerful than that of a 'normal' person, shall this be considered as an enhancement? Should technical

---

[43] Warwick 2003, p 131. This merging of human and machine might go on in the direction of merging humans to some extent by connecting their nervous systems through electronic channels (cf. also below). This can question the 'boundaries' of the concept of a person; cf. also Jaquet-Chiffelle 2006, Koops and Jaquet-Chiffelle 2008.

[44] Warwick 2003, p 134.

[45] Clearly, one could start here discussing cyborg morals, ethics and the like. For this, we refer to Warwick 2003 as a starting point for further discussion and literature.

[46] For aspects concerning social acceptance see Sect. 11.2.5.

progress for instance be restricted to only allow restoration which does not go beyond 'usual' capabilities?[47] A side question would then be how to measure this? Some researchers have made radical statements in this respect: 'From a cybernetic viewpoint, the boundaries between humans and machines become almost inconsequential.'[48]

11.3 Opinion 20 on Ethical Aspects of ICT Implants in the Human Body

In their 2005 Opinion 20,[49] the European Group on Ethics in Science and New Technologies (EGE) discussed a series of principles that are at stake in the case of 'human implants,' some of which have already been discussed in the legal chapter. Hereunder, we briefly summarise the principles they consider pertinent:

- Human dignity, meaning that a person should be able to operate in self-determination in a free society;
- Human inviolability, meaning that (parts of) the human body should not be commodified;
- Privacy and data protection, meaning that 'human implants' should not convey personal data without adequate protection;
- The Precautionary principle, meaning that in cases of scientific uncertainty the introduction of specific technologies that generate unknown risk, require precautionary measures such as innovative research to investigate potential risk;
- Data minimisation, purpose specification, proportionality and relevance, meaning that in as far as the data generated by the usage of human implants are personal data the rules of data protection must be applied;
- Human autonomy, meaning that implants may never be used to manipulate individual citizens in a way that refutes their self-determination in a free society.

The conclusion of the EGE is that the existence of serious but uncertain risks requires the application of the principle of precaution, with special attention to the difference between active and passive implants, reversible and irreversible implants and between offline and online implants.

As for purpose specification the EGE notes that this mandates a distinction between medical and non-medical applications, while as for the data minimisation principle it notes that implants should be replaced by less invasive tools as far as possible.

In the opinion of the EGE the proportionality principle rules out the lawfulness of implants if used exclusively to facilitate access to public premises. The integrity

---

[47] See also the discussion in the following sections.

[48] Warwick 2003, p 131.

[49] Rodotà and Capurro 2005.

of the human body rules out that a data subject's consent is sufficient grounds for using human implants, while the human dignity principle would rule out usage of human implants that turns the human body into an object that can be manipulated remotely and/or turn it into a mere source of information.

The EGE also noted a shift from designing human implants for the purpose of observation (monitoring, diagnosis) to modification of human beings (enhancement, the advent of hybrids). This modification is deemed ethically relevant because the usage of online implants may turn individual persons into networked individuals over whom a certain amount of remote control might be exercised.

The EGE thus takes a clear and perhaps controversial stand on a number of issues, for instance being very strict about the extent to which consent could qualify as a sufficient legal ground for implantation. This concerns Article 7 of the Data Protection Directive, that sums up the legal grounds for the processing of personal data, starting with the consent of the data subject. The EGE seems of the opinion that Article 8 of the Convention of Human Rights overrules such consent in the specific case of human implants, due to the invasive nature of this technology and its potential violation of human dignity. Similarly, the EGE finds that the even in the case of consent, the proportionality principle of Article 6 of Data Protection Directive would rule out implants if used to monitor or authorise access to public premises. From the perspective of democracy and the Rule of Law one can readily agree that requiring a person to accept implants in order to have easy access to a public premise is very problematic. The extensive profiling and surveillance that would become possible if—for the sake of easy access—citizens would consent to implants, seems to rule out such usage as indeed disproportional.

11.4 Ethical Implications of ICT Implants for Democracy and the Rule of Law

11.4.1 Introduction

The contribution of the EGE is salient, especially with regard to the distinction they make between active/passive, reversible/irreversible and offline/online implants. In this section we will add some crucial issues that relate to the wider implications of human implants, by looking into the ethically relevant consequences of four types of implants: those that aim to restore or repair human capabilities (Sect. 11.4.2), those that aim to enable monitoring of biological conditions (Sect. 11.4.3), those that aim to enhance human capabilities (Sect. 11.4.4) and finally those that aim to identify a human person (Sect. 11.4.6).[50] In Sect. 11.4.7 we will also discuss human implants as an enabling technology for (group) profiling, a subject hardly

---

[50] Cf. Hansson 2005 which aims to provide a more or less systematic overview of what is at stake.

touched upon in the Opinion. The relevant implications that we investigate are related to some of the central tenets of constitutional democracy and we will summarise them in the concluding section (11.4.8).[51]

11.4.2 Restoration of Human Capabilities

The borderline between restoration of human capabilities and their enhancement will shift once specific types of enhancement become the norm. This in itself will be a consequence of the availability of these technologies, raising the question of consent as well as distributive fairness. If a certain enhancement becomes the norm a person may no longer feel free to decide against it, because she will probably be disadvantaged if she rejects it.[52] If certain enhancements are relatively expensive they may not be covered by medical insurance, thus creating or reinforcing the gap between those that can and those that cannot afford them.

The restoration of human capabilities, such as in the case of cardiovascular pacers, cochlear implants, deep brain stimulation for Parkinson's disease or an insulin pump, may have ethical implications in as far as they:

1. reduce ethical dilemmas around donations and transplants. Implanting technological devices instead of body parts of other humans or animals may resolve some of the issues related to the need for donated body organs. Organ donation requires invasive interventions in the body of others and impact decisions on the threshold of life and death. The availability of artificial transplants will reduce the chance that another body will be used as a resource[53];
2. increase the scope of end of life decisions, since survival may depend on the implantation, raising the issue of whether and when a device can be turned off when the patient is unconscious;
3. raise distributive issues because insurance companies may not compensate expensive 'repair-implants,' creating or reinforcing the gap between those that can and those that cannot afford them;
4. blur the border between repair and enhancement, raising issues as to the difference between normalcy and disease, as discussed above;
5. induce personality change, especially in the case that implants are used that affect the brain or the central nervous system. Just like medicines used to reduce depression, psychosis or bipolar disease, the sense of self of a person may be seriously affected while a person may also not have any control over the dynamics of this change;

---

[51] Cf. Hildebrandt and Gutwirth 2008, Chaps. 14 and 15.

[52] Cf. Garreau, Chap. 2 'Be All You Can Be.'

[53] For an intriguing narrative about creating and using a set of human beings as a resource of 'fresh organs' we refer to Kazuo Ishiguro's novel *Never Let Me Go* (2005).

6. produce cultural effects because certain conditions, e.g. deafness, may become curable, while the Deaf World (claiming to speak for a linguistic and cultural minority) does not consider deafness a disability to be 'cured';
7. require involuntary interventions in the case that a patient is unconscious;
8. impact privacy, for instance when they have been tagged with an RFID-tag or other devices that allow others to monitor their condition (see Sect. 11.4.3); and
9. enable discrimination, for instance when they have been tagged with an RFID-tag or enhanced with other devices that can identify a person as having a prosthesis.

11.4.3 Monitoring of Biological Conditions and Processes

The monitoring of biological conditions and processes concerns the usage of implants for a diagnosis of biological state of affairs, such as biosensors like Micro Electro-Mechanical Systems (MEMSs) that monitor inaccessible parts of the body and collect data like blood pressure or glucose levels. The same devices can be extended for smart drug delivery or for alerting medical assistance in the case of risk.[54] These monitoring devices partly evoke the same issues as those related above. They may decrease or change the scope and nature of end of life dilemmas in as far as they provide early warning systems that allow more timely interventions; they will raise distributive issues in as far as they are considered as part of preventive medicine that is not covered by insurance companies. These two implications will combine when those who can afford implantations will survive acute health threats, while for those not 'monitored,' help could arrive too late. Also, those who value their privacy may be at a disadvantage when they reject real-time online monitoring via such devices. For this reason standard biological monitoring implants may become the rule, turning consent into a farce (insurance companies may refuse to pay unless one is enhanced with a monitoring implant).

The data collected by these implants will be part of the patient's electronic file, allowing refined profiling and—depending on who has access to these files—extensive social sorting by insurance companies and (potential) employers. This could enable refined—but perhaps unjustified—price-discrimination.

Justice authorities may also be interested in screening such data, and in using implantable devices to control offenders convicted for sexual abuse while they are on parole. MEMS implants could also be used to monitor psychiatric patients, automatically providing smart drugs delivery, enforcing total compliance.

Biological monitoring devices can also be used for enhancement, see Sect. 11.4.4.

---

[54] These devices are in still in development, see for general information: http://www.devicelink. com/mpmn/archive/07/07/015.html and http://www.memsinvestorjournal.com/2006/08/mems_packaging_.html, and e.g. on drug delivery: http://www.technologyreview.com/Biztech/19784/ and on MEMS used to 'train' cyborg-insects: http://www.technovelgy.com/ct/Science-Fiction-News.asp?NewsNum=571.

11.4.4 Enhancement

Cosmetic surgery and neuropharmacology have already faced us with dilemmas arising from human enhancement, especially with regard to the dynamic borders between what is 'normal' and what is 'deviant.' In the case of cosmetics, deviance is usually connected with being less beautiful or attractive, while in the case of neuropharmacology deviance can for instance be connected with functioning at a lower than average level within one's professional life. In a competitive market the incentive to 'enhance' oneself may be so strong that it becomes difficult to understand such enhancement in terms of self-determination or consent. Usage of (ICT) implants for the enhancement of human memory, vision, auditory perception, alertness or other human capabilities, may have further ethical implications, partly reinforcing issues already in play with existing 'therapies' for enhancement. They will again raise distributive issues, depending on whether insurance companies will pay for them. Once certain devices are widely used they may create a new default position which will blur the border between repair and enhancement, as already discussed above. Enhanced vision, augmented auditory capabilities, reduced need for sleep and increased resistance to stress will induce a change in human identity, and may raise the question of what it is to be human. In as far as some will and others will not enhance themselves, the introduction of these technologies will produce cultural effects and a new segmentation of society (the enhanced and the 'ordinary').[55] In as far as the enhancement depends on online connections, privacy as well as autonomy will be at stake: who will have access to the data exchanged and who will be in control of the online connectivity? The enhanced may find ways to discriminate those less—or not—enhanced, and they may also be capable of protecting their privacy as well as prevent social sorting by outsmarting monitoring devices to their own advantage.

In the case of neural interfaces enabling a direct link between two or more human brains or between brain(s) and machines, entirely new issues arise. Such direct 'cyberthinking' could create cyborgs (human–machine hybrids) that plug into and out of others. This raises the question of what it is to be human in an even more radical manner, as well as questions such as whether such developments are desirable and who will decide on the design and introduction of these technologies (computer scientists, cybernetic experts, commercial enterprise or democratic publics). As for constitutional democracy, we may also have to reflect on whether cyborgs are legal subjects in their own right, capable of bringing about legal consequence (entering into contractual obligation, being liable for harm caused), and competent to vote and partake in public debate. Due to the fact that a person could plug in and out of direct neurological interfaces,[56] the issue of causality will become even more complex: which consequences should be attributed to which

---

[55] Warwick 2003.

[56] For first steps in this direction see for example Warwick's experiments with a 100-electrode array implanted in the nerves of his arm, see above and Gasson et al. 2005.

legal person?[57] A person could switch between being unplugged and being plugged into one or more other persons and/or smart machines. Can we still consider this person to be the same person, and if not, how do all these different cyborgs relate to the unplugged person?

11.4.5 The Issue of (Group) Profiling

Profiling refers to the construction and applications of inferred profiles, based on knowledge discovery in databases (KDD), data mining, machine learning or neural networking. The analysis of 'big data' or 'big number crunching' (Ayres 2007), made possible by increasing computer power leads to entirely different implications than mere identification or data aggregation.

As has been noted earlier in the context of RFID systems and Ambient Intelligence the emphasis on data minimisation seems not wholly adequate in the case of refined (group) profiling.[58] In as far as this concerns group profiles built on anonymous data, the protection or hiding of personal data may not be effective.

First, in the case of Ambient Intelligence and less integrated smart applications, hiding one's data could make the environment less smart. In the case of medical applications this would involve taking substantial risks. To protect one's privacy in a situation where one actually wishes to share one's data in order to benefit from diagnostic tools, risk-monitoring or biological enhancement, a less reductive conception of privacy is needed. In that case privacy is not a matter of non-disclosure of personal data, but a matter of transparency. We follow Agre and Rotenberg's definition of the right to privacy as 'freedom from unreasonable constraints on the building of one's identity.'[59] In as far as the implantation of artificial devices into human beings allows for the mining of knowledge that is relevant for the patient or enhanced 'hybrid', this knowledge should be made accessible and 'readable' for the person it concerns. Only in that case will the person be capable of making informed decisions about 'building her identity', e.g. taking certain health risks, dealing with insurance companies, changing one's lifestyle, negotiating with commercial enterprise about access to her personal profiles.

Second, data minimisation may not be effective because profiles are often constructed on the basis of massive amounts of data, mined from other people. This means that hiding one's data does not prevent profiles from being constructed, while one may still be matched with such profiles on the basis of a minimal disclosure of one's data. Group profiling on the basis of data mined from artificial implants could provide crucial epidemiological information about

---

[57] See Jaquet-Chiffelle 2006, Koops and Jaquet-Chiffelle 2008.

[58] Hildebrandt and Meints 2006, Hildebrandt and Koops 2007.

[59] Agre and Rotenberg 2001, p 7.

diseases and/or the impact of enhancement on bodily functions. Hiding such data would not really protect anybody, since the interest in those data is not an interest in the data of an individual person but an interest in the aggregate, statistical level of analysis. Blocking data would restrict scientific research and reduce the possibility to evaluate the long-term impact of artificial implants.

However, this does not mean that data should be shared indiscriminately. It rather means that profiling generates altogether different ethical dilemmas, related to human autonomy and fair treatment. Having shared one's data, one needs to be aware of the knowledge that may be inferred from them, making *transparency of data processing* the hallmark of 'due processing,' requiring adequate communication about the implications of profiling. This should raise the awareness how profiling allows for refined social sorting, price discrimination and manipulation. For this reason the fair information principles should be extended to include the principle of minimum knowledge asymmetry, requiring those that have implants and/or are hybrids to have adequate access to both the logic that rules decisions taken about them and—of course—the nature and consequences of these decisions. To make such a principle of minimum knowledge asymmetry operational we have detected an urgent need for legal and technological transparency enhancing tools (TETs).[60]

11.4.6 Conclusions on Ethical Implications for Democracy and the Rule of Law

By investigating the ethical implications of human implants for democracy and the Rule of Law we may avoid some of the pitfalls of methodological individualism that haunts much of applied ethics. Indeed, this will shift the focus from individual choice for its own sake to individual autonomy as a precondition of a viable public debate, while recognising that personal autonomy cannot be taken for granted as it depends on the way a society has been structured. Society and individual are not seen as opposites but as mutually constitutive. From the perspective of democracy and the Rule of Law, the following ethical values are involved: freedom from unjustified constraints (liberty), the freedom to participate in democratic processes and public debate, the freedom to participate on an equal footing in the economic market (non-discrimination, distributive justice), due process, privacy and identity (legal subjectivity).

ICT implants may impact ethical issues surrounding the transplantation and donation of human organs, the need for involuntary interventions, end of life decisions and the borderline between repair and enhancement. This will involve the freedom from unreasonable constraints that is a core tenet of constitutional democracy. Due to the fact that ICT implants enable extensive profiling, they will

---

[60] See Hildebrandt 2009b.

also facilitate invisible unfair discrimination and segmentation, which will impact the freedom to participate in democratic processes and public debate, and the freedom to participate on an equal footing in the economic market. Market access and participation will also be modified by distributive issues and cultural developments, triggered by ICT implants. Similarly, the invisibility of unfair discrimination and social sorting will disable due process: contesting unfair treatment becomes difficult to the extent that people are not aware of the link with profiling.

Finally, the concepts of legal personhood, privacy and identity will be transformed by the way ICT implants will impact data protection, the borderline between repair and enhancement and also the development of one's personality and cultural belonging.

11.5 Conclusions

While investigating potential ethical implications we stumble on the radical uncertainties of what the future holds for us. In fact the hybridisation of human beings and their ICT implants will raise the question of 'who is us?' Moore's law implies that the relevant developments are dynamic in an exponential way, meaning that we do not know what we do not know as far as the future is concerned.[61] This means that many of the implications we have been assessing will need further exploration. At this point we do not offer answers, rather raise relevant questions.

Human ICT implants will not be restricted to RFID tags. Entire parts of the human body may be replaced and monitored by means of communicating ICT components. Especially within the context of health care many questions are raised, both as to *acceptance of* and *control over* what such implants do to us. In Sect. 11.2 these questions have been articulated in the form of four types of conflicts, namely between freedom and health, supervision and anonymity, self-determination and heteronomy and between security and privacy on the one hand and safety and effectiveness on the other. In the end neural implants and especially attempts at human enhancement will raise the question of the impact on human agency. In Sect. 11.3 we have indicated how the European Group on Ethics of Science and New Technologies (EGE) has taken up these challenges, articulating potential threats to human dignity, human inviolability, privacy, data protection, data minimisation and human autonomy. The EGE has argued for the application of the principle of precaution and highlighted the need to discriminate between different types of implants in different contexts. Section 11.4 fleshed out some of

---

[61] Moore's law, formulated in 1965: 'the complexity of 'minimum cost semiconductor components' doubles once a year, every year, since the first microchip had been produced 6 years before,' see Garreau 2005, at p 49.

the implications of four types of implants for democracy and the rule of law, including the implications of profiling based on the input from ICT implants.

It seems clear that much more work needs to be done to assess the ethical implications of human implants and we like to reference ongoing work that confirms the need for discussion and further research in this area. Several EU funded projects in the sixth and seventh framework programmes have approached this subject or at least parts of it from different perspectives and with different main goals,[62] e.g. ETHICSBOTS,[63] ICTethics,[64] ETICA,[65] EGAIS.[66]

It remains to be seen what impact these projects will have for advances in this difficult subject. One may guess that the issue will probably get more public presence once applications become available that impact the 'average user,' hence when the general public is faced with these new questions and problems and—subsequently—with potential abuse.

---

[62] See at European Commission CORDIS, cf. http://cordis.europa.eu/fp7 and http://cordis.europa.eu/fp6.

[63] ETHICSBOTS (cf. http://ethicbots.na.infn.it, project duration 2005–2008), amongst whose goals was 'to identify and analyse techno-ethical issues concerned with the integration of human beings and artificial entities, [...] to establish a techno-ethically aware community of researchers, [...] on the subject of techno-ethical issues emerging from current investigations on the interaction between biological and artificial (software/hardware) entities.' (from http://ethicbots.na.infn.it/goals.php). ETHICSBOTS produced deliverables, where we would like to especially mention deliverable D5, and therein Sect. 3.2 Ethics of Brain Computer Interface Technologies: '[...] in connection with ICT implants in the human body involving interfacing with information and robotic systems, more extensive studies are recommended, which take as starting point the 2005 EGE opinion on ICT implants in the human body, specialising and problematising the conclusions of that opinion in the context of the Ethicbots domain of investigations.' (Capurro et al. 2008, p 145).

[64] ICTethics (cf. http://www.ictethics.eu, project duration 2009–2012) is a project, which declares as objective that 'There is an urgent need for a systematic analysis of the ESLA aspects of Research in ICT, of the same type as developed by the ESLA (Ethical Social and Legal Aspects) working group on biotechnology, established by the European Commission in the early 1990s.' (http://cordis.europa.eu/search/index.cfm?fuseaction=proj.document&PJ_LANG=EN&PJ_RCN=10728181).

[65] ETICA (http://www.etica-project.eu, project duration 2009–2011) is a project that '[...] will identify emerging Information and Communication Technologies (ICTs) and their potential application areas in order to analyse and evaluate ethical issues arising from these.' (http://www.etica-project.eu/ → 'About the project') This project presents a brief overview of the ethical issues in the so-called field of Neuroelectronics (Rader 2007, p 70ff), a special version of the ICT implants discussed here.

[66] EGAIS (http://www.egais-project.eu, project duration 2009–2012) is a project that 'investigates how ethical reflexivity could be integrated into the research and technology development culture of EU research, so that these considerations become a more natural part of the evaluation and technical development process. It aims to provide guidance to stakeholders in general that will lead towards a coherent and cohesive approach to achieving ethical design outcomes from projects—an approach that begins with the proposal design and continues throughout the project.' (http://www.egais-project.eu/?q=node/4).

We think that further fundamental as well as practice-oriented research has to fill the gaps in this important area. It may very well be that the first field where the problems mentioned above will arise is the domain of profiling, as explained in the previous section. This may get wide public attention since the 'normal user' will be concerned, e.g. the user with an implanted pacemaker.

References

Agre PE, Rotenberg M (2001) Technology an privacy: the new landscape. MIT Press, Cambridge

Ayres I (2007) Super crunchers: why thinking-by-numbers is the new way to be smart. Bantam Books, New York

Berleur J, Brunnstein K (eds) (1996) Ethics of computing—codes, spaces for discussion and law. Chapman & Hall, London

Bynum T (2001) Computer ethics: basic concepts and historical overview. In: Edward Zalta N (ed) The Standford encyclopaedia of philosophy. http://plato.stanford.edu/archives/win2001/entries/ethics-computer/

Bynum T (2006) Flourishing ethics. Ethics Inf Technol 8(4)

Capurro R, Tamburrini G, Weber J (eds) (2008) D5—Ethical issues in Brain Computer Interface Technologies. Ethicbots Consortium 2008. http://ethicbots.na.infn.it/restricted/doc/D5.pdf

Cvrček D, Matyáš V (eds) (2007) D13.1: Identity and impact of privacy enhancing technologies. FIDIS (Future of Identity in the Information Society) Project, 2007. http://www.fidis.net/

Fawcett E (1994) The Toronto Resolution. Account Res 3:69–72

Ferrari A, Coenen Ch, Grunwald A, Sauter A (2010) Animal Enhancement—Neue technische Möglichkeiten und ethische Fragen. Eidgenössische Ethikkommission für die Biotechnologie im Ausserhumanbereich EKAH. http://www.ekah.admin.ch/fileadmin/ekah-dateien/dokumentation/publikationen/EKAH_Animal_Enhancement_Inh_web_V19822.pdf

Fischer-Hübner S, Hedbom H (eds) (2007) D12.3: A holistic privacy framework for RFID applications. FIDIS (Future of Identity in the Information Society) Project, 2007. http://www.fidis.net/

Garreau J (2005) Radical evolution. The promise and peril of enhancing our minds, our bodies—and what it means to be human. Doubleday, New York

Gasson MN, Hutt BD, Goodhew I, Kyberd P, Warwick K (2005) Invasive neural prosthesis for neural signal detection and nerve stimulation. Int J Adapt Control Signal Process 19(5):365–375

Halperin D, Heydt-Benjamin TS, Fu K, Kohno T, Maisel WH (2008) Security and privacy for implantable medical devices. IEEE Pervasive Comput Special Issue Implant Electron 7(1):30–39

Hansson SO (2005) Implant ethics. J Med Ethics 31:519–525

Hildebrandt M (2009a) Who is profiling who: invisible visibility. In: Gutwirth S, Poullet Y, De Hert P, Nouwt S, de Terwangne C (eds) Reinventing data protection? Springer, Dordrecht

Hildebrandt M (ed) (2009b) D7.12: Behavioural biometric profiling and transparency tools. FIDIS (Future of Identity in the Information Society) Project, 2009. http://www.fidis.net/

Hildebrandt M, Gutwirth S (eds) (2008) Profiling the European citizen. Cross-disciplinary perspectives. Springer, Dordrecht

Hildebrandt M, Koops BJ (eds) (2007) D7.9: A vision of ambient law. FIDIS (Future of Identity in the Information Society) Project, 2007. http://www.fidis.net/

Hildebrandt M, Koops B-J (2010) The challenges of Ambient Law and legal protection in the profiling era. Mod Law Rev 73(3):428–460

Hildebrandt M, Meints M (eds) (2006) D7.7: RFID, Profiling and ambient intelligence. FIDIS (Future of Identity in the Information Society) Project, 2006. http://www.fidis.net/

Jaquet-Chiffelle D-O (2006) D2.13: Virtual persons and identities. FIDIS (Future of Identity in the Information Society) Project, 2006. http://www.fidis.net/

Koops BJ, Jaquet-Chiffelle D-O (eds) (2008) D17.2: New (Id)entities and the law: perspectives on legal personhood for non-humans. FIDIS (Future of Identity in the Information Society) Project, 2008. http://www.fidis.net/

Kumpošt M, Matyáš V, Berthold S (eds) (2007) D13.6: Privacy modelling and identity. FIDIS (Future of Identity in the Information Society) Project, 2007. http://www.fidis.net/

Moor JH (2005) Why we need better ethics for emerging technologies. Ethics Inf Technol 7: 111–119

Müller G, Wohlgemuth S (eds) (2007) D14.2: Study on privacy in business processes by identity management. FIDIS (Future of Identity in the Information Society) Project, 2007. http://www.fidis.net/

Nissenbaum H (2004) Privacy as contextual integrity. Wash Law J 79:101–140

Perakslis C, Wolk R (2006) Social acceptance of RFID as a biometric security method. IEEE Technol Soc Mag 25(3):34–42

Rader M (ed) (2007) Evaluation Report, NETICA deliverable. Deliverable D.3.2. http://moriarty.tech.dmu.ac.uk:8080/pebble/repository/files/deliverables/D%203%202%20final.pdf

Rodotà S, Capurro R (eds) (2005) Ethical aspects of ICT implants in the human body. Opinion of the European Group on ethics in science and new technologies to the European Commission, 16th March 2005

Rotter P, Daskala B, Compañó R (2008) RFID implants: opportunities and challenges for identifying people. IEEE Technol Soc Mag

Sotto LJ (2005) An RFID code of conduct. RFID J. http://www.rfidjournal.com/article/view/1624/

Sprokkereef A, Koops B-J (2009) D3.16: Biometrics: PET or PIT? FIDIS (Future of Identity in the Information Society) Project, 2009. http://www.fidis.net/

UK RFID Council (2006) A UK code of practice for the use of radio frequency identification (RFID) in retail outlets, release 1.0

Venkatasubramanian K, Gupta S (2007) Security for pervasive healthcare. In: Xiao Y (ed) Security in distributed, grid, mobile, and pervasive computing. CRC Press, Boca Raton, pp 349–366

Want R (2008) The Bionic Man. IEEE Pervasive Comput 8:2–4

Warwick K (2003) Cyborg morals, cyborg values, cyborg ethics. Ethics Inf Technol 5:131–137

Weber K (2006) The next step: privacy invasions by biometrics and ICT implants. ACM Ubiquity 7(45):1–18

Wood DM (ed) (2006) A report on the surveillance society—for the information commissioner by the surveillance studies network, 2006. http://www.privacyconference2006.co.uk/index.asp?PageID=10

Chapter 12
Pieces of Me: On Identity and Information and Communications Technology Implants

Bibi van den Berg

Abstract Ever since the dawn of mankind, we have used artefacts to extend our physical abilities or to overcome our bodily shortcomings. We use a stick to reach the apples on the highest branch of a tree, or a lever to lift things that are heavier than our own bodies. And we use microscopes and telescopes to see things beyond the natural range of our visual system. Several twentiethcentury philosophers have pointed out that when humans use artefacts and technologies, these often tend to become *extensions* of their bodies: they become incorporated into the user's body schema. Most of us can flawlessly park a car or write with a pen because of this principle. Technologies, although 'other', can become 'part' of a user's bodily repertoire, even if they are not embedded into the human body. At the same time, it is interesting to note that in some cases technologies can be experienced as 'alien', or that they can even lead users to feel 'alienated' from themselves. The former may happen when we are new at using a technology, or when it malfunctions or breaks down. The latter has been shown to occur, for example, in patients who undergo Deep Brain Stimulation. After treatment, these patients sometimes state that they feel estranged from themselves, that they no longer feel they are the same person. In this chapter we use some of the central ideas from philosophy of technology to clarify these two (seemingly contradictory) perspectives.

Dr Bibi van den Berg is assistant professor at eLaw@Leiden.

B. van den Berg (✉)
eLaw Center for Law in the Information Society, Leiden University, Leiden,
The Netherlands
e-mail: b.van.den.berg@law.leidenuniv.nl

M. N. Gasson et al. (eds.), *Human ICT Implants: Technical, Legal and Ethical Considerations*,
Information Technology and Law Series 23, DOI: 10.1007/978-90-6704-870-5_12,
© T.M.C. ASSER PRESS, The Hague, The Netherlands, and the author(s) 2012

Contents

12.1 Theseus' Paradox, or Stability and Change in Relation to Identity

Once upon a time Captain Theseus and his crew took their ship on a long journey. Their journey was so long that over time the wooden planks of which their ship was made started rotting. The crew dutifully replaced a plank every time this happened. After enough time went by all of the original planks had been replaced by new ones. Did this mean that Theseus' ship was not the same ship anymore? Did the ship change its identity since it was no longer made up of the old parts? Was this a new ship entirely or was it still the same ship instead? The deeper question in this philosophical problem, which has come to be known as Theseus' paradox[1] and which was first developed by the Ancient Greek philosopher Plutarch, is: what is identity, really and how does it relate to change, to augmentation and enhancement? This philosophical problem, and its manifestation in Theseus' paradox, is debated until this day.[2]

---

[1] Plutarch 75 A.D., retrieved from http://classics.mit.edu/Plutarch/theseus.html [last accessed on 6 July 2011].

[2] Many arguments have been put forth in favour of, and against, the claim that Theseus' ship is still the same ship. One could argue that a ship (or a person, for that matter) is none other than the totality of its parts, and hence, if one replaces (all of the) parts, it is a different ship. Note that this means that even the replacement of a single plank on the ship entails that it is no longer the same ship. Alternatively, one could argue that a ship is more than the sum of its parts. It is a structure, a set of parts with in a specific configuration and it derives its identity from that configuration. On this line of reasoning, Theseus' ship would still be the same ship, because the structure of the ship has been left unchanged. But if this is true, then when *does* an object change its identity—when is the 'structure' of a thing ever (suddenly?) transformed from one thing into something else? What qualifies as change and when does change lead to a qualitative shift in identity? Finally, one could argue that Theseus' ship is still the same because a ship is more than its parts or its structure—rather, it is an object that is used for certain purposes (sailing the seas), and owned by a specific person (Theseus). Despite the fact that the actual make up of the ship changes, its purposes and its ownership do not. Hence, Theseus' ship is still Theseus' ship in some sense. As with the 'structure argument', this line of reasoning raises the question at what point (or by what

The paradox surrounding the identity of Theseus' ship, of course, also applies to human identity. The cells in our body are replaced every so many years and we change and grow over time; so how does this affect the continuity of our selves? What's more, when we sleep during the night, or are unconscious for some time, for example during an operation, are we still the same person when we wake up? And what is the effect of new experiences, or the addition of new knowledge, or, for that matter, the gradual loss of memories as we grow old, on who we are? It seems that the same thorny questions that apply to gradually replacing the planks of Theseus' ship also pertain to our own identities and the ways in which we conceive of ourselves.

Of course, these issues become even more intricate when we consider the augmentation, extension or alteration of our human bodies by means of information and communication technologies—or ICT implants. If we add one implant, or enhance a single part of our bodies with a technological artefact–say, a pacemaker or a cochlear implant (see also Chap. 4 of this book)—it is obvious that this does not change a person's identity to such an extent that we would say (s)he is no longer the same person, although technically speaking (no pun intended) (s)he is not precisely the same any more. However, what happens if we replace more than one part, or ultimately maybe even (almost) all parts? At what point in time do we stop speaking of the 'same' person, or do we not stop speaking of the 'same' person at all?

This chapter examines the use of technological artefacts–including but not limited to ICT implants–to alter, improve or replace human capabilities and functions, and the effects this may have on identity and self-perception. Two general themes, which on the surface appear to be quite contradictory, will be discussed: first, the fact that when humans use artefacts to extend or improve their physical capabilities, these artefacts tend to become *incorporated* into their so-called 'body schema'. This entails that they can use technologies *as if* they are parts of their bodies, and hence these technologies may come to feel *as if* they are part of their (bodily) identity. Second, we will look at the fact that some technologies, either embedded into or worn on the outside of the body, on some level always remain *alien* to that body. What is more, research reveals some of these technologies even lead to a sense of alienation in the experience of the body, or of the user's sense of self.

12.2 Incorporating Artefacts

The making and use of tools goes back to the earliest days of mankind. Mankind has used tools to, for example, extend our bodily functions and to remedy bodily shortcomings. Cavemen used clubs to increase the strength of their arms' sway and

---

(Footnote 2 continued)

qualification) a ship (or an identity in general) *would* change. Do identities only change when purposes or ownership change?

spears to extend the reach and deadliness of their thrust, and hence were able to kill animals larger, stronger and more dangerous than themselves. They used the same simple tools we still use to reach things outside our physical range—for example, sticks to reach the fruit at the top of a tree, or levers to lift things that were too heavy to lift directly. Over the centuries we have invented more and more of these kinds of tools to increase our abilities and make up for what we physically lack. We now use telescopes and microscopes to see things that are far beyond our ordinary range of vision; we use books and computers to commit ideas to an eternal (or at least more durable) external memory and we use trains, cars and airplanes to cover distances we could not cover with only our unaided bodies. These inventions have come to fruition in a very short space of time at that. Moreover, we also use tools to restore physical abilities we may have had at one time, yet have lost along the way, or to repair damage that results from disease. Common examples include the use of glasses, hearing aids and pacemakers. Many of the latest developments in the field of ICT implants fall in this category as well: cochlear implants, retinal implants and the devices used in deep brain stimulation (DBS) are examples in case.

In the early twentieth century several Continental philosophers, including Maurice Merleau-Ponty[3] and Martin Heidegger,[4] studied the role of tool use in everyday experience and the effects of these tools on the experience of the human body. Their findings have remained significant until this day, and, what is more, as this chapter will show, have a renewed relevance in light of the advent of technologies for human enhancement, including ICT implants.

Both Merleau-Ponty[5] and Heidegger[6] concluded that when we use tools to conduct some activity in the world, then these tools themselves have a way of *disappearing* from our direct experience. When I use a hammer to put a nail in the wall, I am not focusing my attention on the hammer itself–on the way I hold it or on the angle at which I aim it at the nail, or on the position of the hammer's shaft and head. Instead, my attention is focused on the nail and the wall. The hammer becomes an extension of my body, as it were, and the tool itself thus removes itself from my direct focus. It appears as though I can use it 'automatically' or even 'effortlessly'. This is what Heidegger calls 'handiness' or 'readiness-to-hand' (zuhandenheit): 'It is characteristic of something ready-to-hand that it withdraws itself from our attention in order to be used.'[7]

In *Phenomenology of Perception*, Merleau-Ponty conducted an extensive study on human perception and on the experience of our human bodies in relation to the environments that surround us. In this study the notion of a '*body schema*' plays a central role. Merleau-Ponty wrote:

---

[3] Merleau-Ponty 1962.

[4] Heidegger 2000.

[5] Merleau-Ponty 1962.

[6] Heidegger 2000.

[7] Verbeek 2005, p 124.

> ...my whole body for me is not an assemblage of organs juxtaposed in space. I am in undivided possession of it and I know where each of my limbs is through a body schema in which all are included.[8]

A body schema is a mental image that we have of our own bodies, representing our body parts, enabling us to 'know' that body and to 'know' how our bodies relate to the space that surrounds them. Thanks to our body schema we are able to sense that a doorpost is too low and that we must duck in order to fit through it, whether an object is within or outside of our reach in relation to the length of our arms, and how deep we have to crouch to sit down precisely on the seat of a chair.

What is interesting about some tools, says Merleau-Ponty, is that these can become *incorporated* into our body schema.[9] As eloquently stated by Andy Clark:

> It is a common observation [...] that the use of simple tools can lead to alterations in [our] local sense of embodiment. Fluently using a stick, we feel as if we are touching the world at the end of the stick, not [...] as if we are touching the stick with our hand. The stick [...] is in some way incorporated, and the overall effect seems more like bringing a temporary whole new-agent circuit into being rather than simply exploiting the stick as a helpful prop or tool...[10]

Compare this to the idea of readiness-to-hand and we see that not only does the tool disappear from our direct attention, but what is more, it comes to feel as an extension of our own bodies. As if it is a part of our bodies, even though it is not in a literal sense. Tools can thus come to seem like parts of our selves. As Don Ihde has suggested, this entails that the:

> experience of one's 'body image' is not fixed but malleably extendable and/or reducible in terms of the material or technological mediations that may be embodied.[11]

Merleau-Ponty provides a number of different examples in order to explain this phenomenon. First, a woman with a feather on her hat somehow 'knows' the distance between the feather and a doorframe—or more precisely, between her body+the-hat-with-the-feather and the doorframe. By incorporating the feather into her body schema she can 'keep a safe distance between the feather in her hat and things which might break it off. She feels where the feather is just as we feel where our hand is.'[12] A second example is that of driving a car. Most of us, most of the time, are quite adept at driving a car, even through narrow roads and can park it in tight-fitting spaces without bumping into other cars. Merleau-Ponty has suggested that this is so because through experience the car comes to feel as an extension of our own bodies and we can accurately determine where it begins and ends because it has come to be a part of our body schema. The third example he discusses is that of a blind man using a stick to navigate the world. Merleau-Ponty writes:

---

[8] Merleau-Ponty 1962, pp 112–113.

[9] Merleau-Ponty 1962, Chap. 3.

[10] Clark 2008, p 31.

[11] Ihde 1990, p 74.

[12] Merleau-Ponty 1962, p 165.

> The blind man's stick has ceased to be an object for him, and is no longer perceived for
> itself; its point has become an area of sensitivity, extending the scope and active radius of
> touch, and providing a parallel to sight.[13]

The blind man has incorporated his stick to such a degree that he can actually sense the world around him through its ending—he has extended his sense of touch and found a viable alternative to sight.[14] Peter-Paul Verbeek explains that 'the intentional relation between human beings and world is thus, as it were, extended or stretched out through artifacts.'[15] The blind man's stick does not simply mediate between him and the world around him–it also shapes the way in which the blind man 'sees' his world. It becomes part of his perceptual repertoire, and hence part of who the blind man is.

All of this, of course, also applies to (some of) the tools, especially the perceptual ones, that we *do* incorporate into bodies, artefacts we use for human enhancement. Clark describes the case of the Australian performance artist Stelarc, who 'routinely deploys a 'third hand', a mechanical actuator controlled by Stelarc's brain through commands to muscle sites on his legs and abdomen.'[16] Stelarc suggests that:

> after some years of practice and performance, he no longer feels as if he has to actively
> control the third hand to achieve his goals. It has become 'transparent equipment' […],
> something through which Stelarc […] can act on the world without first willing an action on
> anything else. In this respect, it now functions much as his biological hands and arms, serving
> his goals without (generally) being itself an object of conscious thought or effortful control.[17]

When Ihde summarises the role and workings of tools and incorporation he states that embodiment relations to technological artefacts entail a 'doubled desire':

> …on one side, […] a wish for total transparency, total embodiment, for the technology to
> truly 'become me'. Were this possible, it would be equivalent to there being no tech-
> nology, for total transparency would be my body and senses […] The other side is the
> desire to have the power, the transformation that the technology makes available. Only by
> using the technology is my bodily power enhanced and magnified by speed, through
> distance, or by any of the other ways in which technologies change my capacities.[18]

Tools, instruments and other artefacts thus are used to extend our capacities and overcome our bodily shortcomings. At the same time once they become part of our body schema, part of our bodily abilities, they tend to act as if they are part of our self-perception, part of how we conceive ourselves and who we think we are.

---

[13] Merleau-Ponty 1962, p 165.

[14] Also see Don Ihde's discussion of the 'embodiment relation' (Ihde 1990, pp 72–80), and Peter-Paul Verbeek's analysis thereof (Verbeek 2005, pp 125–126).

[15] Verbeek 2005, p 125.

[16] Clark 2008, p 33.

[17] Clark 2008, p 33.

[18] Ihde 1990, p 75, emphasis in the original.

We extend our human abilities and powers by means of a wide variety of technological artefacts, and these artefacts, in turn, sometimes come to seem like a part of us, of our way of experiencing the world.

12.3 Alien Implants

In the previous section, we discussed how technological artefacts can at times become incorporated into the body schema and hence may alter the experience of identity. However, quite the opposite may sometimes be true as well: in some situations using a technological artefact can feel quite *alien* and unnatural. In this section we will look at how this works in more detail.

Quite an extensive line of research exists on the 'alienness'[19] of technology in relation to our human bodies and the experience of identity. Interestingly, alienation in relation to (implant) technologies can take place on two levels, or rather, two different meanings of the term alien are used in research into this topic. First of all, the *object* implanted into, or attached to, the body can remain alien; it is strange or not quite a part of one's self(-perception). Second, while the object that is implanted may come of feel like a 'real' part of the person, the individual$_{+implant}$ can come to feel *alienated from him/herself*. In this case, it is not the object that feels alien but rather a person's experience of his or her own identity. Both forms of alienation will be discussed in turn.

12.3.1 This Piece of Me Is Not a Piece of Me

When using technologies, be they for human enhancement or simply as tools to perform a certain task, these technologies may give us an experience of something that is alien, that is not quite right or not quite a 'natural' part of our bodies. Generally speaking, this happens in two situations:

---

[19] Note that terms such as 'alienness' or 'alienation' are far from neutral, if only for the politicised (read: Marxist) ring they may carry. Moreover, as Petran Kockelkoren rightly remarks, speaking of 'alienation' in relation to technology seems to suggest that, somehow, somewhere, there is a 'natural world' out there or at least a non-technologically mediated, original way of relating to the world and that if we were to not use technologies, we could return to this 'natural world'. However, this depiction of things is false. Our perceptions are always mediated (see, for example, Plessner 1975). Kockelkoren writes: '...*in philosophy of mediation there is no natural substratum to fall back on.* [There is no] *unspoilt, primeval state...*' (Kockelkoren 2003, p 34). Despite these disclaimers, it is still worthwhile to investigate the ways in which individuals can come to feel estranged from themselves, or at least can come to consider themselves at some distance from their original conception(s) of themselves, through the use of artefacts or technologies. This is the interpretation of 'alienness' and 'alienation' that will be studied here.

1. when we are (relatively) new at using the artefact, or
2. when the artefact malfunctions or breaks down.

In the first section of this chapter I discussed Merleau-Ponty's examples of situations in which tools or artefacts were incorporated into the individual's body schema—the woman with her feathered hat, the blind man and his stick, and the person driving and parking a car. However, this incorporation has not emerged overnight. To the contrary, incorporating technologies into our body schemas takes a lot of time and practice—we must *learn* how to drive a car, not just to master the machinery, but also to gradually become aware of its size and space, so that we may come to navigate it through traffic without accidents. Similarly, when wearing her feathered hat for the first time the woman in Merleau-Ponty's example probably brushed the feather against every single doorpost she passed. Only over time will she have learnt to 'feel' exactly where the feather ends. Finally, the blind man surely could not 'sense' the world at the end of his stick without good practice, without gaining extensive experience in its use.

When human beings first start using new technologies they are often quite helpless with them and may experience them as strange, alien tools. Clark points out that there is a difference between *using a tool*, and *incorporating* it into one's body schema. The latter only takes place after long practice; the former occurs when we first *begin* to work with the tool. In that case, when performing a task with it, we must constantly (also) focus on the tool itself, 'by roughly representing the tool and its features and powers (e.g., its length) and calculating effective uses accordingly.'[20] Only once we master the use of a technology to a sufficient degree can we let the tool itself slip to the background of our experience—what Heidegger called 'readiness-to-hand', or also 'transparent equipment'.[21] Until that time, the technology is far from transparent. It requires attention and deliberation to be used properly.

On a cultural level we can also conclude that the introduction of technologies requires a process of adaptation and learning before we are well adjusted enough to use them effortlessly. Kockelkoren calls this the 'decentring' effect of new technologies.[22] He uses the example of the introduction of the train to explain how decentring works. Shortly after the introduction of the first trains a wave of different forms of 'railway sickness'[23] (of which railway spine' has become the most famous) swept over the Western world. Remarkably, after a few decades this disease

---

[20] Heidegger 2000, p 98, also see Clark 2008, p 10.

[21] Clark 2008, p 10, also see Ihde 1990, pp 72–80.

[22] Kockelkoren 2003.

[23] Kockelkoren lists a number of different symptoms, such as '*eye infections and diminution of vision*, […] *miscarriages, blockages of the urinary tract, and haemorrhage*'. Soon thereafter a list of mental disorders was added, ranging from '*mental disturbance*' and '*delirium furiosum*' to '*siderodromophobia*', which Kockelkoren describes as '*the general disorientation accompanied with physical discomfort that seems to have affected the first rail passengers en masse*'. (Kockelkoren 2003, pp 15–16).

disappeared altogether. The emergence and decline of this disease is explained by the fact that experiencing travel on board a train was entirely novel for the early train travellers. It could not be compared to any of their past experiences of travel, either on foot, on a horse or on wheels (carts, coaches etcetera). Kockelkoren writes:

> What is near flashes by, what is further away seems to revolve on its axis as soon as you stare at it. All the while, the body remains motionless. It took a while to appropriate the new experience, but after a few decades the disorientation and the rail sickness it produced disappeared.[24]

The disappearance of this disease, Kockelkoren explains, is the result of a process of what he calls 'recentring': the new technology becomes domesticated and loses its novelty. Thus, a new equilibrium is reached until the next new technology comes along. Both on an individual level and on a cultural level the introduction of new technologies may lead to a (temporary) sense of alienation, then. This alienation is overcome by gaining experience with the technology—by getting used to being on a train as a new means of transport, or by using a hammer often enough so that it gradually becomes 'ready-to-hand'.

But the introduction of new technologies, or the early stages of using a new technology, are not the only situations in which artefacts may feel 'alien'. The same applies to artefacts that break down or malfunction. As we have seen 'readiness-to-hand' occurs when one is using a tool in a fluent, effortless manner. But Heidegger also points out that when tools break down or do not function properly, they suddenly *do* become the focus of our attention, and we may perceive them to be quite alien objects.[25] If the handle of the hammer that I use to put a nail in the wall breaks, my attention shifts from the work at hand to the broken tool and it can only return to the original task when the hammer is repaired or replaced.[26] Heidegger calls this phase being 'present-*at*-hand' (*vorhanden*).

Note that when a hammer breaks while putting a nail in the wall, this is inconvenient. But the malfunctioning itself does not have a strong (physical, emotional or psychological) effect on my experience of my own body. After all, the hammer is not 'attached' to me and does not affect my bodily functioning in any direct way. In the case of technological aides implanted into the body (such as, for example, a pacemaker), malfunctioning or breaking down—and hence becoming present-at-hand— obviously has more serious consequences: it may cause serious physical problems (in some cases even lead to a cessation of life) and/or lead to emotional and psychological strain. The thrust of Heidegger's argument is thus even stronger in the case of technologies that are attached to or implanted into the body: when an implant works well, a user may feel so comfortable with it, that (s)he may at times even forget that it is there (readiness-to-hand). When it malfunctions, however, we may experience a sense of the alienness of (or alienation towards) an implanted object that is even stronger than when the object is not part of the human body.

---

[24]   Kockelkoren 2003, p 17.

[25]   cf. Kockelkoren 2003 pp 17–18.

[26]   Also see Kockelkoren 2003 pp 17–18.

Novelty at using technologies and malfunctioning (implant) technologies can lead to a sense of alienation towards the objects themselves. But that is not the only form of alienation that exists in relation to technological artefacts, and especially ICT implants. In some cases, implants can also lead users to experience a sense of alienation *from themselves*. This is what we will turn to next.

12.3.2 *I Am Longer Myself*

When artefacts come to be embedded into the body, this entails that users must somehow find a way to incorporate these artefacts into their sense of (bodily) identity. They must *identify* with these artefacts in such a way that their sense of (enhanced, repaired, alternated) self can find a new balance, in which the implant is somehow integrated. In many cases, this is quite unproblematic. However, research also reveals that there are cases in which individuals find it difficult to strike a new balance in their self-perception, and become alienated from themselves. This can take place either in relation to their *individual* identity or in relation to their *collective* or *group* identity. To explain these to levels, I will discuss two examples:

1. patients who have undergone DBS sometimes appear to experience a loss of self-identification on an individual level, and
2. members of the deaf community, especially in the United States (US) and Australia,[27] tend to resist cochlear implants or other technological 'fixes' to overcome their deafness, because they feel these remedies disregard their *collective* identity as a community of deaf people.

Let us begin by discussing alienation of self on an individual level. In recent years an ethical debate has emerged over the use of DBS which is used, for example, to reduce the effects of Parkinson's disease.[28] While this treatment turns out to be quite effective in remedying some of the physical effects of Parkinson's disease (including, for example, loss of motor skills), at the same time it has become apparent that the treatment may also have an effect on patients' identity and self-experience.

Witt et al. note that after undergoing treatment with DBS patients oftentimes find it 'alarming' to see that the use of this technology also has effects on their experience of self.[29] To underpin their claims, the authors refer back to an empirical study conducted by Schüpbach et al.[30] In this study, the researchers

---

[27] cf. http://www.deafau.org.au/info/policy_cochlear.php.

[28] cf. Kraemer 2011; Schermer 2011; Rabins et al. 2009.

[29] Witt et al. 2011, p 1.

[30] Schüpbach et al. 2006.

conducted semi-structured interviews with 29 patients[31] who had undergone treatment with DBS to combat motor disability caused by Parkinson's disease. As said, research in the past had shown that DBS leads to 'excellent motor outcome' and that in the vast majority of cases neurosurgery also lead 'to an overall improvement in mood, anxiety, and quality of life.'[32] However, in the interviews conducted by Schüpbach and her team, patients appeared to suffer from different sorts of socio-psychological 'side-effects' to the treatment, ranging from an experienced loss of aim in life, (severe) marital and relational problems, and the disappearance of a desire to have or hold a job. What was most striking (in the context of this chapter) were the team's findings in relation to identity and self-perception:

> Nineteen (66%) out of 29 patients expressed a feeling of strangeness and unfamiliarity with themselves after surgery.[33]

Patients said things like:

> I don't feel like myself any more [or] I haven't found myself again after the operation,[34] *and* I don't recognize myself anymore.[35]

Two-thirds of all patients in this study thus experienced a negative effect on their identities after undergoing this treatment–they felt alienated from themselves.

Twenty percent of the patients involved in the study further stated that they also perceived the implant itself as alien; that they were very aware that an electronic device had been implanted into their brains; and that this gave them an altered body image. A few patients even translated their experience of an alien object inside their bodies to a sense of alienation from themselves. They described themselves in phrases such as, 'I feel like a robot' or 'I feel like an electronic doll.'[36]

Other researchers have interpreted the findings of this study in less pessimistic terms. For example, Felicitas Kraemer has suggested that while some patients do indeed experience a sense of alienation after undergoing DBS treatment, others feel *more* 'like themselves' than they did before the treatment.[37] Whether or not patients will experience the implant, and their own identities as (partially) shaped by the technology, as an enrichment or and estrangement, varies. Hence, no straightforward answers in the ethical debate on DBS's effects on identity can be given. Kraemer has suggested that:

---

[31] Although the sample of this study is small, it is one of the few studies conducted in this field and hence the article is considered a landmark work, to which almost all of the other authors cited in this section refer.

[32] Schüpbach et al. 2006, p 1815.

[33] Schüpbach et al. 2006, p 1813.

[34] Schüpbach et al. 2006, p 1813.

[35] Schüpbach et al. 2006, p 1812.

[36] Schüpbach et al. 2006, p 1813.

[37] Kraemer 2011, p 4.

...the experience of alienation and authenticity varies from patient to patient with DBS. For some, alienation can be brought about by neurointerventions because patients no longer feel like themselves. But, on the other hand, it seems alienation can also be cured by DBS as other patients experience their state of mind as authentic under treatment and retrospectively regard their former lives without stimulation as alienated.[38]

Kraemer goes on to note that:

...some reports suggest that some DBS patients feel that they have found their 'real', 'better' or even 'ideal selves' under treatment. This suggests that, under therapy, they have become how they always wanted to be and call this their 'real selves'.[39]

It is unclear whether the effects on identity that patients describe in studies such as those by Schüpbach and her team are caused by the technological artefact that is implanted into their brain, or by other factors, such as the disappearance of symptoms that patients have often lived with for many years. Schüpbach et al. acknowledge this in their conclusion:

[neurosurgery] can result in poor adjustment of the patient to his or her personal, family, and socio-professional life. Whether this is a purely reactive response to a new situation or whether it is caused by an effect of [Deep Brain] stimulation on behavior, or both, remains to be elucidated.[40]

What is important for the discussion in this chapter, however, is that empirical research has revealed that treatment with DBS can lead some patients (although not all) to feel a sense of alienation from themselves, and that DBS can, at least in some cases, be 'disruptive of [patients'] personal narrative identity.'[41] Some patients may experience a negative impact on their *individual* experience of identity.

However, aside from our individual identities, human beings also experience one or more senses of *collective* identity or group identity. We identify with the (members of the) social circles in which we participate—think of the members of one's household, one's family and friends, one's colleagues or people from church or a sports club–and even value and judge ourselves through the eyes of these important others.[42]

Could this sense of identity also be affected through implanting technologies into, or attaching them onto, the body? One example in the affirmative comes from Benford and Malartre's book called *Beyond humans: Living with robots and cyborgs*.[43] In this book the authors describe a clear case of 'techno-resistance', viz. that surrounding cochlear implants. The deaf community in the US has come out

---

[38] Kraemer 2011, p 1.

[39] Kraemer 2011, p 4.

[40] Schüpbach et al. 2006, p 1815.

[41] Schermer 2011, p 2.

[42] Also see for example Blumstein 2000; Charon 1989; Goffman 1959; Mead and Morris 1934; Van den Berg 2008 and Van den Berg 2010.

[43] Benford and Malartre 2007, pp 46–47.

with a clear rejection of this type of technology–or any other form of technology that could help deaf people 'overcome' their deafness. The central idea in their argument is that deaf people have their own culture, with their own language, strong senses of identification and belonging. By 'healing' deaf people from their disability through technological means, the basis for these senses of belonging and identification is shattered. The collective, shared identity that members of the deaf community experience comes to an end when one is no longer a member of that community, because the deafness is past. By looking at deafness as something that should be overcome if possible, proponents of cochlear implants disregard the fact that turning a deaf person into a hearing one fundamentally alters their identity (for more discussion on this, please see Chap. 4 of this book).

The same arguments that we encountered in relation to DBS—that some individuals will feel *more* like themselves, rather than less, after treatment with DBS—potentially apply to this discussion as well. Some deaf people would surely feel that after their hearing was repaired through the use of a cochlear implant, they would feel more 'like themselves' than when they were still deaf. And whether or not people feel more or less alienated from themselves after undergoing cochlear implant surgery, again, may not be exclusively related to the implant itself ('I feel alienated from myself because of the object that is implanted into my head'), but may also be caused by all sorts of socio-psychological factors. As a matter of fact, the arguments presented by the deaf community seem to point exactly in this direction: that indeed *social* parameters are more important in protecting the self-perception of deaf people than the actual technical changes brought about by implanting the artefact.

What these two cases reveal, though, is the idea that (some) human beings may come to experience a sense of alienation from themselves through the incorporation of implant technologies, not only on an *individual* level, but also in relation to their *group* identities.

12.4 Concluding Remarks

In this chapter we encountered what seemed like two contradictory perspectives on the effects of technological artefacts (including ICT implants) on human identity. On the one hand, some technologies can come to be understood as extensions of the body, and hence come to be experienced as if they were a part of the user's identity. This occurs when we are experienced in using an artefact, and hence in its use the artefact itself can disappear from our attention. On the other hand, we have discussed cases in which individuals consider technologies as alien or even alienating. This occurs, for example, when we are unaccustomed to using an artefact, or when it breaks down or malfunctions. And when technologies are embedded into the body, for example, to remedy disease, these may also lead to a sense of alienation from oneself, both on an individual and on a collective level— in this case it is not the object that is alien, but rather one's own identity.

How can this apparent contradiction be explained? A first answer to this question is this: in our identification with technological artefacts, apparently there is a distinction between experienced use and novel use. Novel use may lead us to consider artefacts as alien, while experienced use may lead us to identify strongly with them. These two phenomena are not contradictory because they describe different *phases* of use. What is remarkable about their juxtaposition, however, is the fact that the same objects that feel alien at some point in time may become so ingrained in our practices of use at a later point in time that we 'lose ourselves in them' entirely. Apparently, technological artefacts have a way of invoking a strong response in any phase—whether it is the rejection of the new, or the embracing of the familiar.

This duality—with or embrace—can also be found in the second form of alienation discussed in this chapter: that of becoming alienated from oneself. As we have seen in my discussion of both the DBS and cochlear implants, implanting the technology sometimes leads to a sense of losing one's old self, yet at other times may also give people a sense of finally becoming who they always wanted to be. In the former case, the patient rejects the treatment and/or the artefact, and/or the results, while in the latter (s)he embraces all of these. In the former case, alienation is the central theme, yet in the latter, the artefact becomes part of the person's new sense of self. Again, there is not so much of an opposition between these two, but rather a difference in perception. This time, the difference does not revolve around different phases of technology acceptance, but around personal (and cultural?) conceptions of living with an implant and overcoming disease or disability through them.

One conclusion seems clear: whether ICT implants and other artefacts are embraced, and come to be experienced as a 'real' part of individuals' self, or whether they are rejected, experienced as alien or alienating, when looking at the experience of identity and users' self-perceptions, incorporating implants is never a neutral affair.

References

Benford G, Malartre E (2007) Beyond human: living with robots and cyborgs. Forge, New York
Blumstein P (2000) The production of selves in interpersonal relationships. In: Branaman A (ed) Self and society. Blackwell Publishers, Malden, pp 183–198
Charon JM (1989) Symbolic interactionism: an introduction, an interpretation, an integration. Prentice Hall, Englewood Cliffs
Clark A (2008) Supersizing the mind: embodiment, action, and cognitive extension. Oxford University Press, New York
Goffman E (1959) The presentation of self in everyday life. Doubleday, Garden City
Heidegger M (2000) Being and time (trans: Macquarrie J and Robinson E). Blackwell Publishing Ltd, London
Ihde D (1990) Technology and the lifeworld: from garden to earth. Indiana University Press, Bloomington
Kockelkoren P (2003) Technology: art, fairground, and theatre. NAi Publishers, Rotterdam

Kraemer F (2011) Me, myself and my brain implant: deep brain stimulation raises questions of personal authenticity and alienation. Neuroethics, DOI:10.1007/s12152-011-9115-7

Mead GH, Morris CW (1934) Mind, self & society from the standpoint of a social behaviorist. The University of Chicago Press, Chicago

Merleau-Ponty M (1962) Phenomenology of perception. Routledge, London

Plessner H (1975) Die Stufen des Organischen und der Mensch. Walter De Gruyter & Co, Berlin

Rabins P, Appleby BS, Brandt J, Delong MR, Dunn LB, et al (2009) Scientific and ethical issues related to deep brain stimulation for disorders of mood, behavior, and thought. Arch Gen Psychiatry 66(9): 931–937

Schermer M (2011) Ethical issues in deep brain stimulation. Front Integr Neurosci 5:1–6

Schüpbach M, Gargiulo M, Welter ML, Mallet L, Béhar C, Houeto JL, Maltête D, Mesnage V, Agid Y (2006) Neurosurgery in Parkinson disease: a distressed mind in a repaired body? Neurology 66:1811–1816

Van den Berg B (2008) Self, script, and situation: identity in a world of ICTs. In: Fischer-Hübner S, Duquenoy P, Zuccato A, Martucci L (eds) Self, script, and situation: identity in a world of ICTs. Springer, New York, pp 63–77

Van den Berg B (2010) The situated self: identity in a world of ambient intelligence. Wolf Legal Publishers, Nijmegen

Verbeek PP (2005) What things do: philosophical reflections on technology, agency, and design. Pennsylvania State University Press, University Park

Witt K, Kuhn J, Timmermann L, Zurowski M, Woopen C (2011, forthcoming) Deep brain stimulation and the search for identity. Neuroethics. DOI:10.1007/s12152-011-9100-1

Chapter 13
The Societal Reality of That Which Was Once Science Fiction

Diana M. Bowman, Mark N. Gasson and Eleni Kosta

> *Tomorrow's science is today's science fiction.*
> Stephen Hawking (Hawking 1995, p. xiii.)

Abstract While it may well be necessary to acknowledge that one day we may become part machine, at least for medical reasons, it is impossible to know whether wider utopian, dystopian or simply pedestrian predictions of human ICT implants are realisable in the long term. However, as this book has sought to describe, debate now will have an immediate bearing on the real world with conclusions that could affect researchers, manufacturers, social institutions as well as our ideals of freedom and human dignity. Surprisingly, it has taken the wider academic community some time to agree that meaningful discourse on the topic of human implantable ICT technology is of value. As developments in implantable medical technologies point to greater possibilities for human enhancement, this shift in thinking is not too soon in coming. However, even at this early stage of the technology's trajectory, greater regulatory and scientific certainty is required. And herein lays the challenge: regulating emerging risks, including health, privacy and security risks, against the broader public interest without compromising the development of a promising and powerful technology. The foundations for the acceptance of all human ICT implants must be themselves designed and built as a

D. M. Bowman (✉)
Department of Health Management and Policy and the Risk Science Centre,
School of Public Health, The University of Michigan, Ann Arbor, MI, USA
e-mail: dibowman@umich.edu

M. N. Gasson
School of Systems Engineering, University of Reading, Berkshire, UK
e-mail: m.n.gasson@reading.ac.uk

E. Kosta
Faculty of Law, Interdisciplinary Centre for Law and ICT (ICRI), KU Leuven, Belgium
e-mail: eleni.kosta@law.kuleuven.be

M. N. Gasson et al. (eds.), *Human ICT Implants: Technical, Legal and Ethical Considerations*,   175
Information Technology and Law Series 23, DOI: 10.1007/978-90-6704-870-5_13,
© T.M.C. Asser Press, The Hague, The Netherlands, and the author(s) 2012

matter of priority to ensure their acceptance and commercial success. Having described the varied issues surrounding the development and deployment of human ICT implant devices, this chapter provides an overview of this book's key conclusions and contributions.

Contents

Medical implants that address biological defects, human enhancement, bioengineering, cyborgs, organs that have the ability to repair themselves in situ, subdermal implants that allow you to transact with computers, technologies that can be deployed to enhance the quality and quantity of an individual's life, transhunanism are science fiction to many, but reality—or at least moving towards reality—for others. And what will the realm of science fact hold for our society tomorrow?

As this book has sought to demonstrate, that which was considered to be science fiction in the field of human implants by many not so long ago is indeed science fact today. Gasson (Chap. 2), Rotter et al. (Chap. 3) and Tadeusiewicz et al. (Chap. 4), illustrated that ICT implants including, for example, the cochlear implant and the pacemaker, have been routinely inserted into human bodies for more than a generation now. Such procedures are commonplace; they occur every hour of every day and, in doing so, improve (arguably) the quality of an individual's life. But they have not been without their critics, and as highlighted by authors such as Tadeusiewicz et al. (Chap. 4), Koops and Leenes (Chap. 10), Hildebrandt and Anrig (Chap. 11) and van den Berg (Chap. 12), even a technological application that seemingly is life-enhancing has the potential to raise numerous and complex ethical, philosophical and legal questions.

The state of the art continues to evolve. Researchers around the world are currently working on developing, for example, the bionic eye in an attempt to restore functioning vision to millions of people (see Chaps. 4 and 10). It is a project that few imagined would become a reality even just two decades ago. But projects such as the bionic eye illustrate the increasing complexity and sophistication of technology, and the way in which the once unimaginable is slowly becoming possible.

In line with such seismic shifts, we are now seeing—as highlighted by authors such as Rotter et al. (Chaps. 3 and 5), Rotter and Gasson (Chap. 6) and Kosta and Bowman (Chap. 9)—simple passive technologies, such as RFID, being routinely implanted in healthy humans. And while such applications are primarily done with the informed consent of the individual effected, their slow but increasing deployment has been somewhat polemic. Unlike restorative technologies, which are largely accepted on the basis of their health-related function and the strict government regulations and standards that govern their use, RFID implants have been received with much more scepticism. As contributors to this book have sought to articulate, RFID implants are increasingly giving rise to a plethora of legal (see in particular Chaps. 7 and 9), ethical (as discussed by Roosendaal in Chap. 8, and similarly canvassed in Chap. 10–12) and privacy (see Chap. 6) concerns.

This book has sought to identify many of the key issues associated with human ICT implants today. For example, much of the debate surrounding the deployment and use of passive implants for identification relates to the potential loss of privacy that the individual may face due to being identified by others without their consent. Or indeed, without even their awareness. For example, despite the short communication range of today's RFID implants, there is still a risk that they may be misused for the physical tracking of a person. Placing readers at both sides of doorframes would, for example, enable the detection of implants in users' arms and enable the unauthorised capture of the stored information.

In a coercive attack an attacker could, for example, force an authorised user to provide his identification and authentication credentials. Such a security threat may occur for any identification and authentication method, but for RFID implants it carries the risk of physical harm, as an attacker could cause injury by extracting the implant from the victim's body. Therefore, it has been argued that RFID implants may be appropriate for identification of people but, regardless of any future development of technical security solutions, they cannot provide secure authentication.

The basic foundations of advanced ICT implant devices are, however, being developed for therapeutic purposes and have the potential to enhance the quality of life of millions of individuals. It is therefore difficult to mount a convincing argument as to why we should put the brakes on research in this area. Moreover, as this book has sought to illustrate, underpinning much of the current debates is the concept of risk. Risk in terms of potential physical harm and risk in terms of broader societal harms. But this is not in itself new, nor is it unique to human implant technologies as Little[1] has sought to remind us: "governments, regulatory agencies and courts have, of course, been evaluating risks posed by scientific, industrial and technological advances in one form or another since before the industrial revolution". As such, institutions and processes have been developed, and are already in place, to help manage and minimise different forms of risk—that may or may not appear depending on the developmental trajectory of the technology and its acceptance by the market.

That is not to say however, that these applications do not introduce a challenging set of questions that must be grappled with by key stakeholders and society more generally over the coming years. Indeed the increasing commercialisation of human ICT implants has generated debate over the ethical, legal and social aspects of the technology and its products. However, technological advancement is a part of our evolution, and the next step of forming direct bi-directional links between human and machine is moving inexorably closer. It is clearly timely to entertain

---

[1] Little 2001, p. 73.

the idea, and to have debate regarding the potential use of this more advanced technology in individuals with no medically discernible need. A number of wider moral, ethical and legal issues stem from such applications and it is difficult to foresee the social consequences of adoption long term that may fundamentally change our very conception of self and sense of identity. Intervention with regard to the possible negative impacts should clearly take place at an early point such that we are not left relying on purely legal measures. Nevertheless, it has started to become a common understanding that the technology is not *per se* 'negative'.

At a legal and ethical analysis level, as the authors of Chaps. 7–12 have shown, the implantation of ICT devices may challenge the right of bodily integrity for every human being, as a further expression of the right to self-determination. Moreover, the use of human ICT implants will provide the skeleton on which a number of future applications may be developed that will enable the tracking, tracing and profiling of the individual. There is an imperative, we would argue, for governments, policy makers and other key stakeholders to take a proactive approach to addressing such issues. When the stakes are so high, it would be prudent for the dialogue between policy makers and the research community to begin now, so as to ensure that the former is not caught unaware.

The use of human ICT implants, especially in the medical sector, has been most welcome as it has introduced devices such as cardiovascular pacers, cochlear implants, DBS for Parkinson's disease, and insulin pumps. Notwithstanding the positive impact of such devices to the health condition of the patients, the restoration of human capabilities and especially the enhancement of existing ones are not free of ethical issues. Nor will they be. Given the current situation and directions in which the technology is heading, it is not too soon to start real debate on the next wave of applications that will make their way into human beings. This includes, for example, the bionic eye. To this end, the European Group on Ethics in Science and New Technologies have published their opinion on the use of ICT implants and notes that implants, if not used properly, may prove to be a threat to human dignity, by at the very least not respecting an individual's autonomy and rights. Such dangers are already present with current medical ICT implant devices, where even simple basic access control is not implemented.

It is clearly evident that technological innovation and market demand will play significant roles in driving industrial innovation and economic growth in the coming years. The continual convergence of emerging technologies such as biotechnologies, nanotechnologies, and the cognitive sciences, with other, more traditional technological platforms such as ICT, promises a range of new products and applications that will enhance human health and well-being. The development of human ICT implants in particular is one such area, and this is particularly so in relation to therapeutic applications. However, even at this early stage of the technology's trajectory, greater regulatory and scientific certainty is required. And herein lays the challenge: regulating emerging risks, including health, privacy and security risks, against the broader public interest without compromising the development of a promising and powerful technology. The foundations for the acceptance of all human ICT implants must be themselves designed and built as a matter of priority to ensure their acceptance and commercial success.

References

Hawking S (1995) Foreword. In: Krauss Lawrence M (ed) The physics of star trek. HarperCollins, New York, pp xi–xiii
Little G (2004) BSE and the regulation of risk. Mod Law Rev 64(5):730–756

Index